shield country

the life and times of the oldest piece of the planet

Red Deer Press

The Publishers
Red Deer Press
56 Avenue & 32 Street Box 5005
Red Deer Alberta Canada T4N 5H5

Credits
Front cover photograph by John Poirier.
Back cover photographs by Tessa Macintosh.
Cover design by Cat Tail Publishing.
Illustrations by Autumn Downey.
Cartography by Marilyn Croot.
Printed and bound in Canada by Friesens for Red Deer Press.

Acknowledgments
Financial support provided by the Canada Council, the Department of Canadian Heritage and the Alberta Foundation for the Arts, a beneficiary of the Lottery Fund of the Government of Alberta.

COMMITTED TO THE DEVELOPMENT OF CULTURE AND THE ARTS

THE CANADA COUNCIL | LE CONSEIL DES ARTS
FOR THE ARTS | DU CANADA
SINCE 1957 | DEPUIS 1957

Canadian Cataloguing in Publication Data
Bastedo, Jamie, 1955–
Shield country
ISBN 0-88995-191-8
1. Natural history—Canada, Northern.   2. Canadian Shield.   3. Taiga ecology—Canada.   I. Title.
QE185.B37 1999   508.714   C98-911218-7

To my parents,
for the wings and roots
that set me on this path.

Contents

Maps

Acknowledgements

As one reviewer put it, "This is a multidisciplinary book if there ever was one." Hence, I have many people to thank for reviewing bits and pieces of the manuscript, which represent the many angles from which one can try to comprehend this complex and beautiful land.

Starting from the ground up, I am immensely grateful to John Brophy who demonstrated indefatigable patience with my unending questions on geology. Thanks also to George Patterson, Bill Padgham and the late Walter Gibbins for helping me unearth various geological nuances. Chris Hanks, Tom Andrews and Kevin O'Reilly provided significant guidance in the archaeology and historic sections of the book. For help on details of climate and snow ecology, I thank Bill Pruitt and Dennis Malchuk; on soils and permafrost, Charles Tarnocai; on vegetation and habitats, Stan Rowe, Karen Hamre, Robin Reilly and Rosanna Strong; on mammals, Kim Poole, Doug Heard and Bob Bromley; on birds, Bob Bromley, Jim Hines, Bob Ferguson, Jacques Sirois and Mike Fournier; on insects, Steve Smith and Emery Paquin; on fire ecology and fire history, Rick Lannoville and Ray Schmidt; and on "Tomorrow's Landscape," Chris O'Brien and Brenda McNair.

For library assistance in finding those gems that kept my nose deep in the books, I thank Alison Welch, Don Albright and Barb Campbell. The long, somewhat open-ended loans of books from the personal northern collections of Kevin O'Reilly and Mike Pichichero are also much appreciated.

For generous funding assistance throughout various stages of this project I thank the Northwest Territories Arts Council, Ducks Unlimited Canada, the Maclean Foundation, Ecology North, Indian and Northern Affairs' Environmental Action Program, and the Canadian Polar Commission.

This book is written to be accessible to readers from all walks of life who may share only one characteristic: a heartfelt curiosity about the northern Canadian Shield. If, in massaging the science and stories of this land towards this end, I have allowed any inaccuracies to slip by, I must take sole responsibility. I should add, however, that during the prolonged and wide-ranging review of this book, I came to understand the sometimes fuzzy distinction between perceived inaccuracies and polite disagreements.

I am grateful to all of the contributing photographers, particularly John Poirier, who have enlivened my text with their rich images, and to Autumn Downey, who had the uncanny ability to pull dim images out of my head and bring them to full fruition in her exquisite illustrations. For assistance with preliminary cartography, I thank Janet Troje and Karen Pentland.

A colossal thanks to Mike Robinson of the Arctic Institute of North America for believing enough in this book to write me that "I am pleased to inform you..." letter in response to my publication queries. I am especially indebted to Karen McCullough, Ona Stonkus and Jeremy Drought for their ruthless but ever-patient editing of the original manuscript, their design and formatting of the book and their vigilant shepherding of the final pages though the printing process.

I would also like to thank all publishers and authors who granted me permission to use direct quotes from their material. While every effort was made to trace copyright holders and obtain written permission, this was not possible in all cases. For any omissions I take sole responsibility and trust that the endnote citations and corresponding reference list do justice to all relevant sources.

Finally I extend my deepest gratitude to my wife, Brenda Hans, who in so many countless ways made this labour of love come true, and to my two daughters, Jaya and Nimisha, who brought me cookies and hugs when least expected but most needed.

I do dimly perceive that whilst everything around me is ever-changing, ever-dying, there is underlying all that change a living power that is changeless, that holds all together, that creates, dissolves, and re-creates.

— M.K. Gandhi[1]

Introduction

Whatever evaluation we finally make of a stretch of land . . . no matter how profound
or accurate, we will find it inadequate. The land retains an identity of its own, still
deeper and more subtle than we can know.
— Barry Lopez, *Arctic Dreams* [2]

My natural home — and the subject matter of this book — is the taiga shield, a distinct
ecological region that forms the evergreen, granite-studded crown stretching across two-
thirds of North America. This region is defined by the meeting of two of the biggest
physical features on earth, the taiga forest and the Canadian Shield. Taiga, a Russian word
pronounced "TIE-gah," refers to the northern coniferous forest, that "land of little
sticks," which spans from Labrador to Alaska and beyond, from Siberia to Scandinavia.
In northern Canada, much of this forest rests on the Canadian Shield, the geological
nucleus of the continent. [3]

In size, the taiga shield approximates that of western Europe with one-hundredth the
number of people. A satellite view of this region on a winter's night shows tiny, widely
scattered blips of light twinkling against a jet-black background — islands of human
settlement adrift in a sea of subarctic wilderness. On the far western edge of this region
is one of the larger, though still dim, blips: Yellowknife, capital city of the Northwest
Territories and base camp for the writing of this book.

Near the wilderness fringe of Yellowknife is a gravel pit close to my heart. Unofficially
it is known as the "Bristol Pit," named after an airplane parked in the sky beside it — a
navy blue 1935 Bristol Freighter, the first wheeled aircraft ever to land at the North Pole.

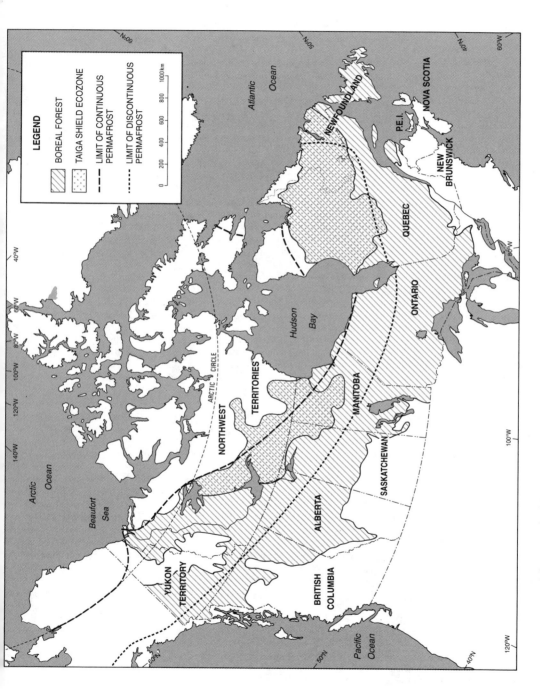

The taiga shield ecozone, defined by the overlap of the taiga forest and Canadian Shield.

Still pointed north, suspended just above the pines in a perpetual landing mode, the plane bears down towards the one and only highway into town. This plane is impossible to miss. At some point, most tourists and probably all locals have visited the Bristol Freighter, a monument paying tribute to our species' bent for rolling back frontiers. On the other hand, the neighbouring gravel pit goes unnoticed by just about everyone.

I consider this pit to be another monument of sorts, worthy of at least equal attention. It was not always so for me. For years I too drove past this lowly pit without giving it a moment's notice. It was a very large, rather dusty hole in the ground, nothing more. Only after I started digging into the scientific literature on this region's fascinating geological history did it dawn on me that such a site was worth visiting.

My first of many pit stops occurred on one clear, balmy evening in August. Risking prosecution — or so the faded sign said — I jumped the wire fence and proceeded down a steep trail into the bowels of the gravel pit. Viewed from the bottom, its high, sloping walls, boulder terraces and craterlike shape gave it the aspect of a majestic Greek amphitheatre badly in need of repair.

The sky that evening was a Himalayan blue. Low-angled sunlight flooded the walls before me with intense golden light. I stood quietly for some time, enjoying the warmth at my back, savouring the stillness of this unusual place and gazing at the seemingly hodgepodge patterns of disintegrated rock. Clearly some drama had occurred here. But over forty years of gravel extraction had messed up much of the story line.

Now risking life, limb and camera, I scrambled up the loose rock for a closer inspection of one of the more well-defined boulder terraces. Here was a thick layer of pink and black and zebra-striped rocks. Many were shaped and sized like watermelons. The story began to make sense as I pondered their forms and stroked their tombstone-smooth surfaces. I found myself imagining the sound of flowing water — writhing torrents of water — in volumes and violence that only a melting continental ice sheet can put out. Like sapphires in a jeweller's tumbling box, these rocks had rolled along the bottom of meltwater streams pouring out from under a glacier over 9,000 years ago.

From my terrace perch I gained a clear view of other multicoloured rock layers deposited by outflow from the receding ice sheet. Some layers were swirly and convoluted, which, seen in cross section, resembled slices from a giant marble cake. These fluid patterns of coarse pebbles echoed the tortuous path once taken by glacial meltwater through ancient rapids now long dried up. Here was a sort of fossilized river.

Glassing around the pit with my binoculars, I noticed an intriguing band of fine, blonde-coloured sediments near its upper rim. I scrambled upwards through space and time to get a better look. Up close, the meaning of these layers became wonderfully obvious. Here was fine, wave-washed sand laid down like a sandwich in hundreds of flat, even beds. I was looking at the distinctive signature of Glacial Lake McConnell, an oceanic megalake that once joined the basins of Lake Athabasca, Great Slave Lake and Great Bear Lake. The story of this phantom water body suddenly came alive before me as I read the lines of its shifting shores.

Moved by a kind of rapture, I turned towards the empty amphitheatre and began applauding, shouting, waving. My yelps of approval and delight brought a small cascade of pebbles down on my head. A band of ravens roosting on the far side of the pit exploded into the air. In these brief moments, bounding up and down the sheer walls of loose sand and stones, I came to a deep visceral appreciation of one short chapter of this landscape's story of creation. The grand performance of an ice age in decline was recorded here plainly for all to see — for all, that is, that have a knowing eye and a curious heart.

That night in the gravel pit something happened to my perception of earthly time. The receding glacier seemed near, so near that, despite the sultry August air, I imagined I could feel subtle wafts of dry, cold wind blowing off the ice's rotting surface. Though the ice sheet disappeared from this land over 8,000 years ago, it seemed like only yesterday, a frozen drop in the fathomless bucket of geological time.

While I wrote this book — utterly steeped in the literature and language of the taiga shield — such irrevocable insights were not uncommon. My field notes record several other odd adventures inspired by an urge to connect all that I had read, heard or dreamed with the concrete reality of this land. For instance, I recall one April morning, thrashing excitedly through leaden hip-deep snow to examine a small patch of freshly exposed Precambrian bedrock, something I had spent most of the winter reading about. The slog was worth it. Just to touch the rocks and examine their jewel-like crystals was a thrill.

Later that year, while researching the plant life of northern wetlands, I made a trip to a black spruce bog to become better acquainted with some species I had paid little attention to before. The turning point in this field trip came when, at the outer limits of the floating bog mat, I suddenly sank up to my chest in wet, soggy peat, bringing the plants I had come to befriend to about the level of my nose.

From all such field trips, I returned sometimes wiser, sometimes confused, but always changed and somehow more alive. The wisdom came from prying deeply into geological or biological secrets and suddenly discovering new meaning and beauty in the land — much like finding the light switches in a darkened art gallery. The confusion came from trying to push back the frontiers of my knowledge about this landscape too far or too fast, forgetting that all of its aspects are not knowable, measurable or necessarily logical. In my quest to understand and explain my natural home, I sometimes forgot to pause and celebrate its mystery.

The primary focus of this book is the western portion of the taiga shield.[4] This area is bounded to the north and east by the tree line, a broad zone of transition between the boreal forest and the treeless arctic barren lands. To the west, it is bounded by the edge of the Canadian Shield, which cuts a clean line south from Great Bear Lake, through the middle of Great Slave Lake and on to the north shore of Lake Athabasca. The denser, more diverse growth of the southern boreal forest marks this area's southern boundary.

Much has been written about this region's human history, about its pivotal role in the northern fur trade, its place in the story of more than one gold rush and its position as a cultural crossroads for the first peoples of this land — the Dene and Inuit. But surprisingly little has been written about the region's unique natural history — its fascinating story of creation that goes back to the dawn of the planet's history, its exposed plains of bald rock that provide an unrivalled showcase of Precambrian geology, its wild rivers and millions of pristine lakes and its status as an ecological crossroads where climates, plants, birds and mammals from two worlds, the Subarctic and the Arctic, overlap in a richly textured wilderness fabric.

Beyond the natural science of this landscape is a unique enchantment or charm, what some would call a "spirit of place." For me, the best way to convey this spirit is through the window of my own subjective experience. As such, this book is part personal journal, drawing on events that illustrate my relationship with the land. It is also part storybook, portraying the land's past, present and future as I see it. It is also part reference book, complete with systematic descriptions of ecological phenomena, an extensive glossary of terms and a detailed index. And finally, it is part field guide, providing sufficient information on the region's geology, plants and animals for you to recognize the main ecological players on this particular northern stage.

In choosing to write a book about the taiga shield, my aim was as much to inspire as to inform. It is written therefore not from the perspective of a dispassionate spectator but a full-fledged participant in the life and times of the land. My wish is that, in musing on the words and images that follow, your own participation in this place, or others like it, is somehow enriched.

An Eagle's View

It is not a land filled with static grandeur, such as the mountains offer. It is rather an unassuming wilderness, a land very content to be unnoticed but a land filled with process and change for those who have the eyes to see. It is at once a gentle yet brutal wilderness that stretches out before us.

— J. David Henry, *Taiga*[5]

There are many ways to befriend a landscape. If you live in Arnprior, Ontario, you can hop in a canoe and drift down the Ottawa River to get a feel for "The Valley." If Moose Jaw, Saskatchewan, is your home, you can launch a pickup truck down a few back country roads to connect with the prairie landscape. If you hang your hat in Pincher Creek, Alberta, a horse will do for a tour of the foothills. Or if it's Hope, British Columbia, you can climb a mountain for an alpine view.

Around Yellowknife, Northwest Territories, there are no lazy rivers down which to drift. Roads are scarce and horses scarcer. And our mountains, all of them, are long gone. To get a feel for this country, it's best to take to the air. . . .

Pointed due north, we taxi down a liquid runway. Thumbs up. Seconds after the floatplane lifts free of the water, the landscape demands my full attention. As Yellowknife shrinks away into a vast sea of subarctic wilderness, the familiar tapestry of the Canadian Shield begins to show off its primary colours — green, grey and blue. The pattern of forest, rock and water repeats itself endlessly in all four directions. From 1,000 metres up, the landscape at first seems bare and monotonous. Yet, ironically, only from the air can I see its astonishing variety.

Like migrating pods of giant whales, rounded bedrock hills rise up into the sunlight, then plunge below a choppy surface of spruce forest and muskeg. These arching dome-shaped outcrops — aptly called "mammillated" by geologists reminded of a female profile — are one of the unmistakable features of shield scenery.

Carving through these domes are long, sloping ridges that end in a sharp drop, often as a vertical cliff more than 100 metres high. From the air I can trace some of these sharp linear ridges or faults clear to the horizon. Forgetting their fascinating geological history for a moment, I press my nose against the plane window and wonder what cosmic surgeon made these unerring slices through the bedrock landscape.

In some places, different rock colours line up neatly along both sides of a fault. Pink granites and greenish-black volcanics are aligned like blankets spread end to end. Elsewhere, rock colours appear stirred together randomly, as in salami, or shuffled, one into another, like a deck of cards. And on every exposed rock face, regardless of shape or colour, are the scars of disintegration and change: folds, fissures, cracks and wrinkles, whole hillsides falling to pieces in huge blocks.

The shapes, patterns and colours of these rocks hold a million mysteries about the planet's history. The oldest known rocks are down there, almost four billion years old. They have seen a lot of action in their time — the roots of ancient mountains, the floor of inland seas, the brunt of ice sheets three kilometres thick. And here they are below the airplane, now supporting the world's youngest forests.

As if to heal this scarred and ancient landscape, ribbons of green seek out the most furrowed and pitted rock surfaces where meagre soils tend to concentrate. From the air, vegetation of the taiga shield often forms a mosaic of relatively distinct plant communities, coexisting side by side with little mingling going on. Dense spruce forests, luxuriant birch groves, muskeg mats and ridge-top benches of pine all know their boundaries. Together they express, with stunning fidelity, the landscape's subtle variations in moisture availability, microclimate and soil conditions.

Fire ignores such boundaries and adds to this mosaic by leaving in its wake a patchwork of forests in various states of renewal. I spot a distant plume of grey smoke on the western horizon, reminding me that fires are as much a part of this northern ecosystem as the trees they burn.

We are not equipped to fight fires. Instead we turn our tail to the smoke and descend a few hundred metres to have a closer look at all these lakes. In the late afternoon sun, their deep blue, jewel-like quality is irresistible to my eyes. Below us are scores of lakes, most of them nameless. Their shifting, glistening surfaces give expression to a marvellous union of sun and wind.

On every exposed rock face are the scars of disintegration and perpetual change. (*John Poirier*)

The pilot's map shows hundreds, maybe thousands, of lakes within reach of a full tank of gas. Those that are named speak of frontier adventure and hidden riches in the land: Desperation Lake, Trapper Lake and Prosperous Lake. Their names also speak of tranquility, music and the stirring of souls touched by the unique charm of this landscape: Bliss Lake, Prelude Lake, Mystery Lake.

There are great, sweeping lakes with aquamarine water reminiscent of the Caribbean Sea. There are much smaller, tea-coloured lakes — giant puddles really — that sport floating bouquets of a thousand yellow pond lilies. There is everything in between. From up here, some lakes look like cartoon characters, animals or human faces, with the occasional island serving as an eyeball or nose. Again, like jewels, no two lakes have the same cut, nor do the countless rivers and streams that join them.

Some rivers in this country are no more than chains of lakes connected by short, turbulent stretches of white water. The more fortunate rivers find major faults, folds or fractures in the rocks, which provide shortcuts to the next place of rest. Other rivers and small streams follow more tortuous courses, with the path of least resistance taking them on a convoluted journey around unyielding bedrock hills and corrugated mounds of glacial debris.

The amount of water in shield country is almost unbelievable. Just north of Yellowknife, lakes cover almost 30 percent of the landscape — and that's just lakes. My comprehension staggers when I add to this equation all the water flowing unceasingly down the many rivers and streams, all the water locked up in soggy muskeg or frozen ground and, for good measure, all the water in Great Slave Lake, Canada's third largest freshwater body. That's a lot of water.

For purity, this water is unmatched — perhaps in all the world. In fact, water taken from the Yellowknife River is used as an official standard by which to judge drinking water throughout Canada. But, as our plane sweeps low over a shallow noose-shaped pond, I wonder about the purity of some of the smaller, poorly drained lakes. The one below us looks pretty murky. I conclude that it was probably stirred up by a moose just passing through. I crane my neck sideways to catch a last glimpse of the fast receding pond. Sure enough, there's a trail of hoof prints visible in the shallows.

Biologists depend on aircraft to keep tabs on moose, as well as on caribou and migrating ducks, geese and swans. Whether flying for work or pleasure, luck always plays a role in what, if anything, is seen. From the air, I was once lucky enough to spot a school of arctic grayling loafing in a still, clear pool near the tree line. Unfortunately for the fish, we were in a helicopter fully equipped with fishing rods.

Today, as our plane glides above the shore of yet another nameless lake, we spot a large group of arctic terns now on full alert status. The shadow of our plane streaks across their island colony. The terns explode from their roosts. They wheel in graceful arcs above the island, creating an instant aerial shield against predators with a hungry eye for their newly hatched chicks — quick-footed fuzzballs on the run.

Beside a waterfall at the end of the lake we discover a huge bald eagle's nest in a leaning veteran spruce. The tree appears ready to topple over any second under the weight of thirty years' worth of sticks it didn't grow. In the centre of this tangled mass are two flightless eaglets. Heads cocked and defiant, they follow our plane with unwavering stares.

As much as I would like to proceed to the tree line and beyond, the red line on my map tells me that we have reached the northern limit of our tour. I am consoled to know that we will be back this way again soon during our annual fall pilgrimage to the tundra. Already a few lemon-coloured tops of birch and poplar trees reveal hints that fall is near.

I get the pilot's attention and twirl a finger in the air. The plane banks slowly as she turns its nose south towards Great Slave Lake. Halfway home, my reverie on the landscape is interrupted by a sudden starboard lurch of the plane. The pilot gestures for me to look out the opposite window: our moose. This king of the deer family is oblivious to our passage. Submerged to the neck, he munches contentedly on the succulent roots of yellow pond lilies.

The pilot begins our final descent over the Yellowknife River. An endless tapestry of familiar images unfurls below the plane: bald rock outcrops spattered with multicoloured lichens, azure waters dotted with bobbing gulls, spires of spruce punctuating a crystalline sky — for me, these are signatures of shield country.

Near the north shore of Great Slave we spot a bald eagle gliding in tandem with the plane for a few moments as if to guide us home. In these moments with the eagle, the charm and mystery of this country sink into me just a little deeper. I reflect on the time it takes to befriend this sometimes overwhelming landscape.

An occasional bird's-eye view certainly helps. The apparent chaos of bedrock and bush on the ground falls into patterns that begin to make sense from the air. Things seem to fit together in an elemental unity of rock, water, fire and life. Gazing at the stark beauty of my favourite landscape below, I feel I am looking straight down into Nature's face.

The Making of a Landscape

Introduction

The rocks of shield country record a fascinating landscape history that began almost four billion years ago, a mere six hundred million years after the dawning of earthly time. Here is the story of the main actors on this stage — from volcanoes to glaciers, dinosaurs to caribou, spearthrowers to golddiggers.

The Shield is bedrock, the primal stuff.

— Barbara Moon, *The Canadian Shield* [6]

1

Born of Fire:
Shield Country's Earliest Days

In shield country, rock is the ecological bottom line. It determines the lay of the land, the pattern of vegetation, the flow of water. It also dictates patterns of human settlement. Yellowknife owes its existence to gold buried in the rock, a town "where the gold is paved with streets" — a hackneyed saying, but true. Regular explosions reverberating deep below my house attest to the ongoing quest for gold.

Look at a map of the region. A good proportion of the lakes that anyone has bothered to name owe their identity to rocks: Rock Lake, Rocky Lake, Roundrock Lake, Redrock Lake, Rocknest Lake. Many are named after early geologists — rock hounds. They raved about "the great stretches of clean, bare rock . . . providing ideal conditions for the prospector."[7] Early non-native visitors to the region were so preoccupied with rocks that the resident Dogribs gave them the name Kwet'i, meaning "Rock People," a name that remains very much alive today.

For generations, the rock of the Canadian Shield has played on the imagination of those who call it home, those who stay only a while, even those who have never seen it. It has left its imprint on the nation's best literature, art and drama. It's no wonder. Almost two-thirds of the country is Canadian Shield — over 4.8 million square kilometres. As a geological entity, it doesn't stop there. From its northwestern margin, it dips south below the waters of Lake Athabasca, below the boreal forest of northern Alberta, below the prairies. Rocks kindred to those on the north shore of Great Slave Lake peek out again at the bottom of the Grand Canyon — it's the same primal stuff.

This rock forms the so-called nucleus of the North American continent. Other geological structures have taken up position millions of years after the Canadian Shield was formed. The Rockies, for instance, are relative newcomers on the scene, formed a mere 60 million years ago. The Himalayas are barely out of the geological cradle, having risen just 35 million years ago. Volcanic upheavals in the northwestern Canadian Shield fizzled out about 2.5 *billion* years ago. On the scale of geological time, this happened during the very first chapter of the planet's history — the Precambrian era — hence the name "Precambrian Shield."

As monolithic as the shield may seem, it is actually a composite of seven distinct blocks or geological provinces, each having a different geological history, physical characteristics and a clearly defined boundary. Underlying the apparent uniformity of today's western taiga shield are three of those provinces: the Slave province, which spans from the northeast shore of Great Slave Lake to the arctic coast; the Bear province, which nestles against the east shore of Great Bear Lake; and the massive Churchill province, which stretches north, east and south of Great Slave Lake clear to Hudson Bay and Baffin Island.

Oldest of these provinces and the most intensively studied is the Slave geological province. Internationally acclaimed as a showcase of Precambrian geology, the Slave, like all other shield provinces, is made up of three main kinds of rock: igneous, sedimentary and metamorphic.

Perhaps out of a healthy respect for the underworld, geologists have paid tribute to the gods in describing rocks born of fire. "Ignis," the Roman god of fire, is immortalized in the term igneous, referring to two types of rocks forged from superheated magma deep within the earth. Cracks formed during mountain building, faulting and other earth-shaking events provide a pathway for pressurized magma to shoot towards the earth's surface. Volcanic rocks are formed from magma that makes it all the way up to the surface and spills out onto the land as lava. Among geologists around the world, the Slave province is famous for its well-preserved volcanic rock, particularly the greenstones where most of the gold is lodged. Plutonic rocks, named after the Roman gatekeeper of hell, are formed from magma that solidifies deep below the earth's surface. The giant blobs of granite found throughout the Slave province came into being as plutons, once buried kilometres down beneath ancient mountains.

So what happened to the mountains? Many of them ended up as sedimentary rock. Over millions of years, all mountains eventually disintegrate through the relentless processes of weathering and erosion. Lofty peaks are unroofed and converted to rock fragments, sand or minute particles of silt. Brought low by water, wind and ice, these sediments are deposited on river deltas and quiet ocean bottoms, where they build up,

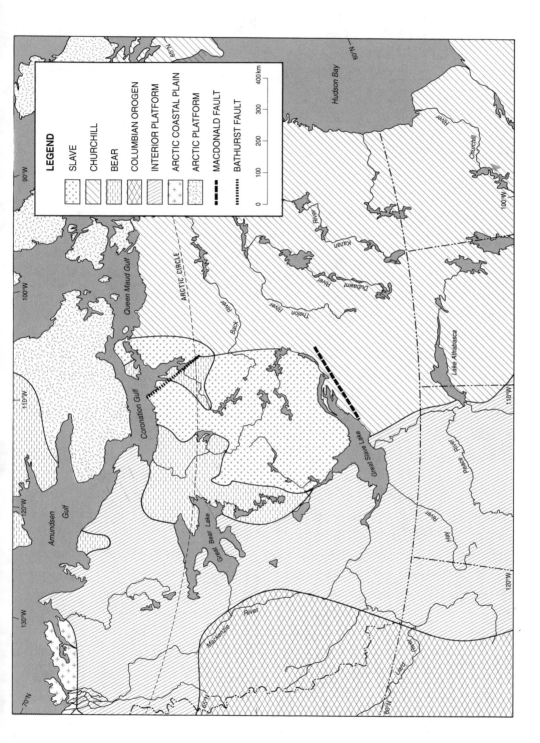

Geological provinces and major faults.

layer by layer, often to depths of several kilometres. As pressures increase, things heat up and the layers become cemented together to form solid rock. Compared to other shield provinces, the Slave has an unusual abundance of sedimentary rocks. Like the volcanic greenstones, some are well preserved. Others have been stretched, squashed or warped beyond recognition. They've been metamorphosed.

Rocks tend to go through an identity crisis when buried beneath huge sediment piles, engulfed by lava flows, invaded by plutons or generally bent out of shape. Due to the heat and pressure generated in these situations, they are baked like bricks in a kiln and emerge as metamorphic rock. For instance, if conditions are right, shale turns into slate, limestone into marble and granite into gneiss (pronounced "nice"). The most common metamorphic rock found stirred into most of the Canadian Shield is gneiss.

The intricate folds, faults and banded layers in rocks made of metamorphic gneiss are a record of the many contortions they must have gone through during their long and complex history. The oldest known rocks are gneiss, found on a small unnamed island on an unnamed lake in the Acasta River basin 300 kilometres northwest of Yellowknife. Although their "parentage," as geologists say, is unknown, their age can be determined with remarkable precision. By using a Sensitive High-mass Resolution Ion Micro Probe — otherwise known as a SHRIMP — geologists have logged these rocks in at 3,962,000,000 years old.[8] Only rocks that have dropped in from outer space have surpassed that record.

With a proven history of almost four billion years, rocks of the Slave province date back nearly to the very dawn of the planet. Just another billion years or so earlier, the earth, according to popular theory, was little more than a thick swirl of cosmic dust and hot gases. As particles slammed into each other, tremendous heat was released, converting the original mass into a molten blob. Heat was gradually radiated into space and heavier elements like nickel and iron settled towards the centre to form the planet's dense, liquid core. Lighter elements rose to the surface and solidified to form a thin crust consisting of granitic lumps floating like icebergs on a continuous film of heavier basalt.

Innumerable volcanic vents opened as a result of further crustal cooling and regular bombardment by asteroids and other interstellar flotsam. Lava piled up, forming the nuclei of future continents. At the same time, volatile gases were expelled, forming a steamy primitive atmosphere of, among other things, water vapour, ammonia and sulphur dioxide. The stage was set for some pretty serious acid rain. With further cooling, the rains did come — for millions of years. Pools became lakes, lakes became oceans. Sticking their grey hills up above these oceans was a flotilla of granitic protocontinents, one of which would become the Slave geological province.

Somewhere in that primordial mass of granite was the rock that someday, about four billion years later, would be probed by a SHRIMP. The Acasta gneiss from the western margin of the Slave province is the nearest thing anyone can find of the earth's original crust. This in itself is exciting. But more than that, it holds the key to understanding just what was going on in the earliest days of the Slave province after the crust was formed.

The unrelenting erosive action of water, flowing or frozen, has removed several kilometres of rock from the original Precambrian landscape, reducing it to the undulating rock plain that we see today. *(John Poirier)*

In its subtle textures and skewed layers, this crystalline time capsule provides a record of major geological events starting 3.96 billion years ago. The Acasta gneiss could be to Slave geologists what the Rosetta stone was to early interpreters of Egyptian hieroglyphics. Unfortunately there's not much of it around.

Over the next two billion years, new rocks continued to pile up on top of the original crust. Volcanoes blew their stacks and disappeared. Some rocks were folded savagely and, having nowhere to go but up, fed rising mountain chains. Others were forced downwards, sinking into the depths, where they melted, becoming magma once again, only to be recycled later into a new generation of rocks.

Just as volcanic activity began to die down around a billion years ago, the Slave province was shaken by a series of geological spasms that created its remarkable faults. Rock families were torn apart, sliding or dipping past one another — sometimes for several kilometres — adding further confusion to the already jumbled landscape. Never stopping its work for a moment, erosion has since flattened virtually everything, leaving behind a rock plain in which original relationships are obscured or downright obliterated.

What were those geological "events" of the Precambrian era that Slave geologists speak of with as much reverence as uncertainty? What bizarre planetary processes caused the major episodes of volcanism, folding and faulting? There are, it seems, as many explanations as there are geologists. It's not that geologists by nature enjoy dickering. It is just that none of them were around in those days. All these stupendous earth-shaping events went on unnoticed.

Federal geologists based in Yellowknife have made finding, dating and interpreting the world's oldest rocks a "long-term project." I wish them luck. Much more is known about the early history of the moon.[9] Crustal activity petered out up there around 3.8 billion years ago. Since then, nothing much has happened. Meanwhile, down on the Slave province, violent upheavals continued for almost another 2.5 billion years, messing up or destroying most of the earliest record in the rocks. It may take generations of geologists to unravel the main events that shaped the Slave, let alone to figure out what caused them.

Until then, my favourite version of the story will remain the one told to me by John Brophy, one of those federal geologists looking for answers. Having waded over my head through several other versions of the story, I found his enthusiasm for the subject and his slightly wry approach refreshing. On his desk, beside a bag of chocolate fudge cookies, sat a stately looking chunk of Acasta gneiss. To help interpret its folded layers, one side of it had been cut by a saw and polished like a tombstone. As Brophy casually took me back a few billion years — as only a geologist can — I occasionally glanced with genuine awe at the zebra-coloured rock. Was it listening?

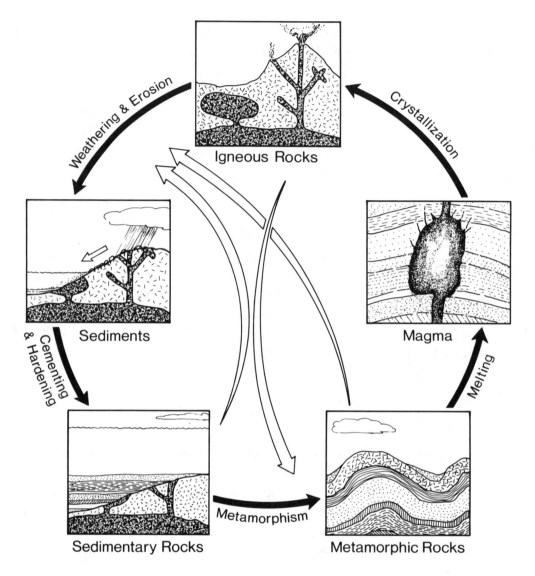

The rock cycle — like everything else in nature, rocks are constantly being created and destroyed, changing into something else. In this slow but inexorable recycling process, rocks take on many forms as they circulate within the earth's crust.

"Naturally if you've got a piece of real estate four billion years old trundling around the globe," Brophy mused, "it's going to be affected by all the things going on deeper in that globe. Every little event is going to leave its imprint on it. By the very virtue that rocks in the Slave are so old, they probably bear the scars of every major event that took place on the entire globe in the last four billion years."

Brophy paused for a moment to calculate how much trundling a protocontinent could do in just one billion years. "Okay, with the crustal plate moving at, let's say, 5 centimetres per year, that's 50,000 kilometres in a billion. Just imagine how often plates have gone around banging into each other, maybe joining for a while then breaking up, meeting again someplace else. They've probably gone several times around the world. Therefore it's pretty safe to conclude that mountains have been built up everywhere."

I gazed out the window of Brophy's office at the low, knobby hills around Yellowknife. He made his point clear. "Look at those flat, deformed rocks. You know that there was a mountain there."

Hungry for proof, I picked up the Acasta gneiss for a closer look. Clearly this rock had a story to tell. Brophy's index finger came into view. "You see a texture like this? A fabric like that? Probably a major deformational event. And look, you see how this band overprints that one? Probably another." Although nodding, I did not see half of what he was pointing to. I did, however, believe him. Like a fortune-teller in reverse, Brophy gazed into the crystal mass, recounting several violent episodes in the story of this ancient landscape. "You can see even more with a microscope," he added. More nods. My vision was clouded by images of folding rock, splitting earthquakes and surging lava, telltale signs of a land in labour — giving birth to mountains.

Only mountain building, it seems, could have subjected the Acasta gneiss to the heat and pressure needed to explain these contorted patterns. Where the mountains came from is another story, and again, there are many versions. One version has mountain ranges springing up like bow waves along the leading edge of the Slave as it was borne along on its circumnavigation of the planet. Another version has an asteroid punching a hole in the Slave province, creating an enduring hot spot that caused mountains to ruck up as the plate moved over it. Yet another version has the Slave province slamming into a younger protocontinent to the east, now called the Bear province, then ricocheting off, only to collide into its massive eastern neighbour, the ancestor of today's Churchill province. Lots of mountain building potential there.

All of the above could have happened — or none of it. There are enough deformities in the rock record of the Slave province to suggest several rounds of mountain building or other so-called deformational events, "four or five at least," in Brophy's opinion. Each event could have had a different cause. Many of the explanations hinge on the mechanism for moving crustal plates around the globe. The way many believe it works today is that, like froth on a boiling pot of soup, the earth's continents drift about on its surface driven by swirling convection currents of hot magma below. Did they move in the same way and at the same speeds as they do now? In this realm of early plate tectonics, geologists are split right in two.

No one can say with certainty how things worked on or below the earth's surface billions of years ago. The debate goes on in the arcane journals of geological science. Some geologists believe that the process of tectonic movement as we now know it has been going on pretty much since day one. Others hold the opposite view, believing that the monstrous scale of mountain building in the planet's earliest days points to a different magnitude or kind of plate activity that we can't even imagine. According to Brophy, "Authors writing on this topic in the same journal can be absolutely convinced either way." Having no interest or credentials for such debate, I suggested to him that we had reached a major threshold of knowledge in the history of the Slave geological province. It was his turn to nod.

After stepping out of Brophy's office and back into the late twentieth century, I lingered for a while in the adjacent corridor. Covering almost every square centimetre of its walls were tacked-up displays of weird shapes and rainbow colours reminiscent of a day-care centre after a long but creative afternoon. Instead of pumpkins or red-nosed reindeer conjured up by toddlers, I was looking at globular plutons, curving volcanic belts and amoeba-shaped sedimentary basins all neatly mapped out by the Geological Survey of Canada. On the apparent chaos of shield country's rock surface, order had been imposed. At least so it seemed at first glance. Among the hodgepodge of bizarre shapes and colours indicating different rock types, there was a liberal sprinkling of dashed and dotted lines, indicating uncertainty, and the occasional question mark, indicating basically an educated guess.

Some of the most intensively studied rocks of the Canadian Shield, and hence most accurately mapped, are those embracing Yellowknife Bay, poised on the southwest edge of the Slave province. In fact, few areas on earth have been so closely scrutinized. Lured by the promise of gold, prospectors and geologists have been scrambling over and under this area for almost sixty years. During this time, its depths have been poked full of drill holes and mine shafts, its surface photographed from just about every angle and altitude, its chemistry analyzed in innumerable laboratories and its magnetic anatomy dissected by sensitive geophysical instruments. All this work has yielded not only a lot of gold but also a detailed record of the Precambrian turbulence that created this piece of the planet.

2

Pillows, Plutons and Pockets of Gold: The Yellowknife Supergroup

Bare rock outcrops within a few minutes' walk from Yellowknife's City Hall preserve evidence of all of the major geological processes that shaped the Canadian Shield: volcanism, sedimentation, metamorphism and faulting. The mayor's desk chair itself is perched over the likely site of a volcanic fissure, defunct for 2.5 billion years. Better exposures of bedrock this old and this well preserved are hard to find, making Yellowknife an international showcase of Precambrian geology.

Perhaps no one has a more intimate knowledge of these rocks than Bill Padgham, another federal geologist with 20 years' experience in the region. Like Brophy, he has a special fascination with rocks older than 2.5 billion years, formed during the early Precambrian era known as the Archean eon, which literally means "the most ancient time." In a technical field guide prepared for fellow geologists, Padgham waxes eloquent in his introduction of the Yellowknife landscape, conveying obvious excitement, affection and unabashed praise — enough to bring any closet rock hound out of the closet.

> The magnificent exposures in the Yellowknife area, and the excellent preservation of volcanic and sedimentary features provide unrivalled opportunity for study of volcano-sedimentary features in Archean rocks. The belt is one of the best known in the world and detailed study continues to produce new data, new explanations and new insights into Archean evolution.[10]

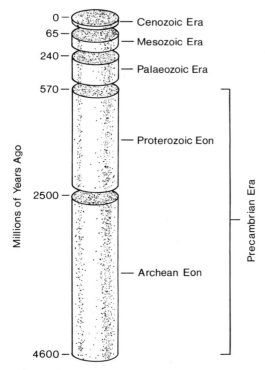

The geological time scale. Shield country's rocks record almost all but the first 600 million years of the planet's history.

The Yellowknife area displays most of the traits of a typical Archean terrain. It is dominated by granitic rocks originally born as igneous plutons, huge pockets of red-hot magma that solidified deep beneath the earth's evolving crust to form roots of the rising Archean mountains. Set in this granitoid sea are smaller areas of rock that formed on top of the crust — the so-called Yellowknife Supergroup, consisting of a mixture of well-preserved volcanic and sedimentary rocks.

The original mineral structure of the Supergroup rocks has changed little since they first solidified eons ago. These rocks have suffered only slight metamorphism since the days of their creation between 2.70 and 2.65 billion years ago. The older volcanic components have a slight greenish tinge, best appreciated from a high hill in summer when the evening sun brings out the full warmth of their colour. These are the celebrated greenstones, the centre of Yellowknife's geological showcase.

The Acasta gneiss is the only evidence we have of the Slave province's original crust that was laid down around 4 billion years ago. The greenstones are all that's left of what came next: a colossal outpouring of lava that piled up on top of the original crust for over 50 million years, reaching depths of 20 kilometres or more. It's no surprise that there is little remaining evidence of the original crust.

The ultimate cause of this massive and prolonged flood of lava has long been a subject of debate among geologists. Explanations are based on information that is, according to Padgham, "both limited and contradictory."[11]

One popular theory suggests that the great lava flood was triggered by some new turbulence in the swirling red-hot mantle about 6 kilometres below what is now Yellowknife Bay. Deep-seated currents of magma flowing away from each other began stretching the original crust, causing it to thin and eventually fracture along a linear zone of rifting. Lava began spilling out onto the original basement crust, first in fits and starts through isolated vents, then more steadily, in unimaginable volumes as the rift zone eventually grew to a length of 15 kilometres.[12] The Yellowknife greenstones clearly show that much of this activity happened underwater, below the surface of an ocean that filled a large basin now largely filled with rock.

In sufficient depth of water, submarine volcanoes in Archean times were a relatively subdued affair, spurting out huge volumes of lava without causing a ripple on the surface. Below 300 metres, the pressure of water overhead was so great that volatile gases and steam stayed locked in solution with the molten rock. Emerging from multiple vents in the earth's crust, rivulets of lava oozed lazily down the rising subsea pile, eventually congealing into globular masses of volcanic rock called pillows.

Seen in cross section, like overgrown spaghetti sliced with a fork, volcanic pillows show a distinctive pattern of oval cells of rock that can measure several metres across. Seen intact, as curling blobs on the landscape, they could easily be mistaken for petrified dinosaur dung. Volcanic pillows cover half of the planet's surface. Their formation carries

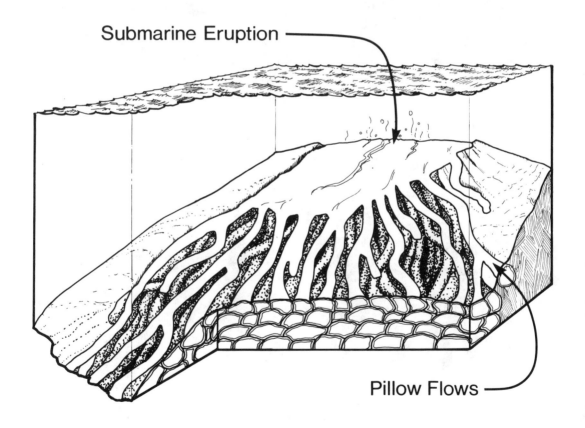

Volcanic pillows, arising from fissures deep beneath a Precambrian sea. Under tremendous water pressure, the lava flowed downslope in a tangled mass of saclike, bulbous lobes.

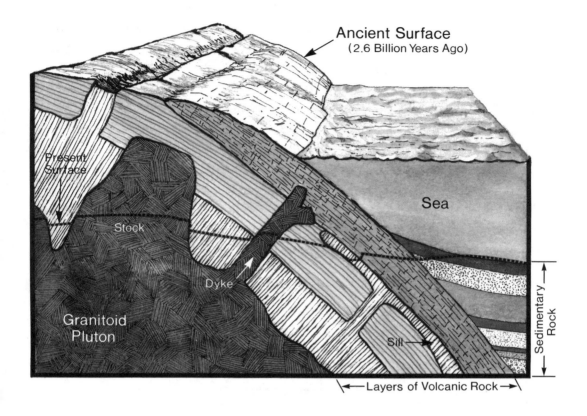

Ancient Surface
(2.6 Billion Years Ago)

Present Surface

Sea

Stock

Dyke

Granitoid
Pluton

Sill

Sedimentary
Rock

←— Layers of Volcanic Rock —→

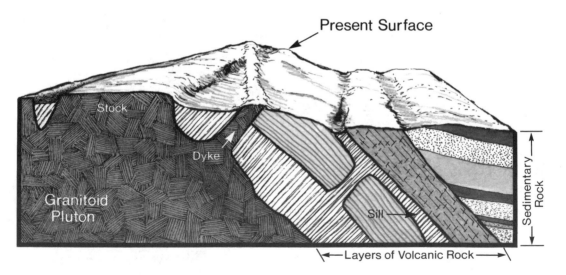

Present Surface

Stock

Dyke

Granitoid
Pluton

Sill

Sedimentary
Rock

←— Layers of Volcanic Rock —→

Cross section of Archean landscape 2.6 billion years ago (top) and today (bottom). Towards the end of the Archean eon, some monstrous spasm below the earth's crust created large plutons that tilted the overlying volcanic pile. Originally intruded deep beneath the surface, these plutons and associated dykes and stocks are now exposed to the light of day after billions of years of weathering and erosion.

on today in such well-trodden places as Hawaii. Yet Yellowknife has become an internationally acclaimed place for geologists to see pillows close up. Scuba tanks or small submarines are usually needed to see pillows elsewhere, since most of them carpet the bottom of present-day oceans.

Glassy shards, ash and small gas bubbles embedded in other Yellowknife greenstones provide a record of more violent eruptions that must have occurred in shallower water. Layer upon layer of lava was laid down until the volcanoes eventually broke through to the water surface, sending rooster tail clouds of steam, ash and lava bombs into the primitive atmosphere.

Having engulfed the floor of an entire ocean basin and risen many thousands of metres towards the surface, this massive pile of volcanic rock must have presented an imposing testimony to the awesome forces that created this piece of the planet. But as true then as now, what goes up must come down. Equally powerful were the forces of destruction that worked away relentlessly at the volcanic pile. Billions of years of weathering and erosion have left us with mere slivers of greenstone rock, scattered vestiges of a landscape once dominated by volcanoes.

While the fresh volcanic surface wore down and the sedimentary rocks built up, cracks and chasms deep below were being intruded by colossal pulses of magma, forming large granitoid plutons. Most impressive of the Yellowknife plutons, for its sheer size, is the Western Plutonic Complex, a vast sea of pinkish rock butting up against the western edge of the Yellowknife greenstones. Beginning around 2.6 billion years ago, at least four different intrusive phases invaded an area 350 kilometres long, 100 kilometres wide and probably 10 or more kilometres deep.

Now exposed to the light of day, this enormous pluton was formed by mighty upward surges of magma, which found cavernous weak spots in the fractured underbelly of the volcanic pile. Today, the greenstone remnants of that pile dip sharply to the east, away from the pluton. Some of them have been flipped right upside down.

Why all this tilting? Because as the pluton formed, the tremendous swamping of magma from below likely triggered the birth of mountains above. The greenstones were lifted up, riding high on the sides of Precambrian mountains with molten hot roots.

Only once in the earth's history were such monster plutons created. They are found in Archean rocks around the world — Australia, Greenland and India. They mark one of the most violent events ever to shake the earth, and one of the least understood. One persuasive theory suggests that, under the growing volcanic and sedimentary piles, part of the original basement crust sank into the red-hot mantle, where it melted to become magma, creating new instability below. Something had to give.

At some explosive stress threshold, huge volumes of stirred-up magma rammed up into the base of the overlying rock, recycling itself as the Western Plutonic Complex. Lucky for us, plutons of this magnitude no longer pound the earth's crust from below. Just why not, nobody knows.

Better understood is the origin of the layered rock lying east of the Yellowknife greenstones. These are the sedimentary rocks of the Burwash Formation, named after Major "Lockie" Burwash, who, in 1934, discovered a spectacular pocket of gold in a quartz vein slicing through the rock along the east shore of Yellowknife Bay. Like the volcanic greenstones, the sedimentary layers have been amazingly well preserved since their creation in Precambrian times. Many are as old as 2.7 billion years, which means that they were forming long before volcanic activity had died down or the giant granitoid plutons had been born.

Fragments of volcanic rock dominate the older sediments, derived both from eruptions and erosion, the latter process having begun on the first day of the great lava flood. The most ancient sedimentary rocks contain scattered granitoid fragments as old as 3.5 billion years that must have eroded from the original basement crust of the earth — chips off the old block, you might say. Younger sedimentary layers show an increasing proportion of granitoid fragments mixed in with the volcanics. Their emergence in the rock record marks the time when nearby plutonic terrains were unroofed and they too began to erode.

Geologists can detect sediment pathways in the rock that indicate from what direction the material came before it settled down on the ocean bottom. Like fossil ripples, these palaeocurrents, meaning "ancient currents," indicate that most of the sediments came from the west — from the Western Plutonic Complex and the volcanic belt on which Yellowknife is perched today.

Just east of Yellowknife is one of the largest sedimentary basins on the Canadian Shield. It measures about 150 by 75 kilometres in total area and is filled with what local geologists call the Burwash sediments, in honour of the major himself. These were deposited in remarkably even beds consisting of fine, dark mudstones and coarser, usually lighter greywackes, which have a rather unscientific pronunciation: it's "gray-wackies." The mudstone layers were laid down in quiescent periods during which minute particles of clay and silt settled gradually, grain by grain, to the ocean bottom over a period of several thousand years. On the other hand, the greywacke layers were laid down more abruptly in a matter of days or hours following violent rainstorms or seismic shocks. These events sent thick slurries of runoff cascading into the ocean, creating underwater floods of chunky sediments.

As these separate layers piled up, they became cemented together under the pressure of the increasing load above, reaching up to 5 kilometres or more in places. This cycle of deposition continued for millions of years, interrupted by occasional rounds of mountain building, intrusion of new plutons or other crustal seizures that flipped, folded and faulted the original sedimentary beds. In many areas within the Burwash Formation, the result of these late Archean catastrophes is startling: layers of rock once laid down like a sandwich now stand upright like a half-squeezed accordion.

It was during these turbulent times that most of the area's gold likely emerged from the fiery depths. Lockie Burwash's discovery along Yellowknife Bay in 1934 was found in a quartz vein that snaked 72 metres up and over the crest of a low hill just in from the shore. This gold was free; that is to say, it was visible to the naked eye — and exceptionally concentrated. A decent ore grade runs about a quarter ounce of gold per ton. The 16 tons of vein material excavated from the Burwash mine yielded a phenomenal 13.6 ounces per ton! Stories, most of them true, about other gold-studded quartz veins fuelled the mining boom that followed. Veins covered with "gold warts as big as a man's thumb" and "Treasure Islands" littered with loose nuggets of gold were among the images that danced in the heads of early prospectors.[13]

Gold-bearing quartz veins are relatively common in the southwestern Slave province — "as common as ships were in the English Channel on D-Day," according to Ray Price, an early Yellowknife historian.[14] Over 100 have been documented in the Burwash sedimentary basin alone. But few have been mined. Quartz veins tend to yield high-grade ore but in low volumes, making gold removal economically unattractive. The mother lode for 80 percent of all gold mines in the region rests in shear zones, an altogether different type of geological structure, but the process that laid down the gold was basically the same.

During late Archean times, about 2.5 billion years ago, the fresh rock surfaces were repeatedly folded, uplifted, compressed and stretched, creating numerous cracks and crevices, big and small. They took many forms, depending on the type of rock involved and the nature of forces at work. Sedimentary rock is relatively plastic and tends to break along isolated cracks, some of which are now filled with quartz. Shear zones were formed in more brittle rock, usually volcanic, where a powerful wrenching in the earth's crust created a large, tangled network of split rock, which, as John Brophy puts it, was "fractured like crazy."

The more massive shear zone systems and some quartz veins are believed to reach down several kilometres below the surface. They formed in deep cracks and fissures that provided channel-ways for the earth's churning interior to blow off steam, to release hot fluids and gases created by deep-seated volcanic events. These hydrothermal solutions contained a steamy broth of dissolved minerals, including quartz, feldspar, iron and, apparently, gold. Pulsing repeatedly upwards, these solutions occasionally encountered some magical combination of physical and chemical conditions — just the right temperature, pressure, acidity and so on — in which gold was no longer comfortable in solution. Escaping from the hot broth, the gold solidified as glittering flecks icing the rock walls. In rare cases — call them nature's accidents — these gold-bearing fluids left behind huge ore bodies, which, after more than two billion years of erosion and some imaginative sleuthing by early geologists, would be dug out of the ground and put back into solution, only to be crystallized again as a shining yellow brick.

A hefty sample of gold dust wrested from the Burwash sediments. *(Busse /NWT Archives)*

By the end of the Archean eon, the main pieces of the Yellowknife Supergroup puzzle were in place. The main period of vulcanism and sedimentation was over and the major rock formations had taken up their present positions: a huge granitoid pluton to the west, the Burwash sediments to the east and at the centre the volcanic greenstones. Most of the gold and other valuable minerals had been laid down in subterranean nooks and crannies. Bathed in the rays of a primeval sun, the land lay bald and lifeless.

But there were stirrings in the sea. Embedded in the late Archean sediments are broken filaments, wormlike casts and clusters of tiny spheres that bear no resemblance to rock crystals of any kind. These are microfossils, precious evidence of the planet's first life. The oldest unequivocal signs of life come from a discovery of stromatolites in the shield rock of western Australia. These toadstool-like colonies of blue-green algae are 3.5 billion years old, almost as old as the Acasta gneiss.[15]

On most of the Canadian Shield, evidence of Archean life is scanty and difficult to interpret. The heat and pressure of billions of years of metamorphism have obliterated most of it. Though still rare, ancient fossils are richer and more diverse in bedrock formed during the next major chapter in the earth's history, the Proterozoic eon. Literally meaning "the period of first life," this was a time when stromatolites ruled the earth, especially between 2 billion and 680 million years ago. Rocks of the East Arm of Great Slave Lake date back to the Proterozoic. Some of the world's largest and best preserved stromatolite fossils are found along its shores.

Here, huge reefs of blue-green algae once flourished in shallow tide pools and sheltered lagoons along the margins of a tropical sea. They formed floating mats up to several metres wide, with trunks of organic material reaching down below the surface. Leaking out of these mats was a gas largely foreign to the atmosphere of this young planet: oxygen.

Like modern plants, the blue-green algae that formed stromatolites were able to perform the miracle of photosynthesis — they used energy from the sun to make sugars out of carbon dioxide and water. Oxygen was released as a kind of exhaust fume. Ironically, blue-green algae cannot tolerate much oxygen. Nonetheless, by the height of their dominion over the planet, they had raised the oxygen level in the atmosphere to around 1 percent, compared with just over 20 percent today.[16] That was enough to begin establishing an ozone layer that could block ultraviolet radiation, thus removing a major obstacle to the proliferation of life on land.

Stromatolites had created a world for which they themselves were remarkably ill suited. But in the process, they had swung open an evolutionary door for radically new forms of life that came to depend on oxygen for survival.

The best way to preserve creatures as fragile as blue-green algae in the fossil record is to quickly smother them with large volumes of sand. Such was the likely fate of the stromatolites now ringing Blanchet Island, Utsingi Point and other fossil showcases along the East Arm. A series of torrential storms might have buried them, or maybe a tidal wave. It easily could have been a tremendous earthquake. Though nothing could match the violence of the Archean eon, these were still turbulent times. Geologically, all was not yet calm.

During the Proterozoic eon, the nucleus of North America was still under construction. The Canadian Shield was still in pieces. One of those pieces, the Slave geological province, had come of age by then and was behaving as a coherent block, basically the same size, structure and shape it is today. But it was still trundling. In the early days of the Proterozoic eon, it was drifting in a southeasterly direction on a collision course towards its final resting place (if anything is "final" in this business). With a long, drawn-out thud that may have lasted a million years or more, the Slave province smacked broadside into the northeast edge of the younger, more massive Churchill province, then ground to a halt around 1.85 billion years ago.[17]

The shock resulting from this collision added a distinctive Proterozoic signature to the Archean landscape. A large area of softer sedimentary rock along the northwest margin of the Churchill province was crumpled like a car bumper, creating the East Arm Fold Belt. At some point, the bottom fell out of this area along two parallel faults, forming a 600-metre chasm, now home to some of the world's most sought-after lake trout. On the southern side of this fault trough is a long, 180-metre-high ridge that slices cleanly through the otherwise rolling landscape for 100 kilometres. This is the great McDonald Fault, which formed where the Slave and Churchill provinces slipped violently past each other in opposite directions.

Similar faults were made along what is now Bathurst Inlet to the north and, later, along the western margin of the Slave, where the Bear province is said to have collided from the rear. Together, these major faults constitute the main suture marks binding the Slave province to the rest of the Canadian Shield.

Few geological events create more stress in the earth's crust than the collision of continental plates. Shock waves rippling from the point of impact caused many kinds of secondary effects besides faulting at the plate margins. These included localized volcanic activity in the form of ribbonlike dykes and small plutonic intrusions, fracturing and folding of rock surfaces and widespread metamorphism. All of these events shook and shaped the Slave province during the Proterozoic eon, events that were likely triggered by the Slave plate bashing into, or being bashed by, its neighbours.

The most visible and perhaps most intriguing secondary effects of these collisions are several large northwesterly trending faults. Paying little heed to variations in topography or rock type, these faults carved a linear gash through the heart of the Slave province. One of the most impressive of these is the West Bay Fault, which now splits Yellowknife in two. If the McDonald and Bathurst faults were the so-called "master faults" created by the Slave-Churchill collision, then the West Bay Fault surely must be a close second. It forms the spine of a 16-kilometre-wide belt of interconnected faults that extends up the Yellowknife River for at least 80 kilometres north of Yellowknife Bay. Within this belt are more than ten major faults, along which rocks on the east side of the fault line have slipped northwards from 1 to 5 kilometres. In the case of the West Bay Fault, this means that the land on which Yellowknife's Old Town sits came to rest after being rafted 5 kilometres north of its former position.

Early geologists found telltale signs of this displacement by tracing the patterns of intrusive dykes that cut across the fault line. These are sheetlike masses of rock that formed when magma was injected into deep-rooted cracks measuring a few centimetres to many metres across. On the surface, dykes typically run in almost straight lines, sometimes for several kilometres. Most were formed a little over 1 billion years ago, long after the original rocks were formed but before the period of intense faulting and displacement. The disjointed paths of prominent dykes on the surface of today's landscape provide the best clues for determining how faulting might have moved things around down below.

This became a topic of hot debate once the legendary Giant Shear Zone was discovered in 1944. The question on everyone's mind was: "Did the fault split this rich ore body and, if so, where is the rest of the gold?" It was the geologist Neil Campbell who cleverly figured out that it lay 5 kilometres to the south along the West Bay Fault. What later became known as the Campbell Shear Zone was once mated to the Giant Shear Zone, long ago, before the Proterozoic traumas that shook the entire Slave province.

Since then, not much has happened to these rocks. The Proterozoic faults represent the final punctuation marks in the development of the Slave geological province. By the close of this eon, all the other pieces of the Canadian Shield were assembled and locked into place. Together they formed a rigid bedrock nucleus around which mountains rose and huge basins filled with sediment. As a new continent was being born, the shield's creation story ended. Here, the forces of destruction — weathering and erosion — have had the upper hand ever since.

On this gradual, inexorable yielding of stone to the elements, Henry David Thoreau gave us this thought:

> The finest workers in stone are not copper or steel tools, but the gentle touches of air and water working at their leisure with a liberal allowance of time.[18]

3

The Ascent of Life:
The Next Half Billion Years

Oddly enough, more papers have been written, more conferences held, more guesses made about the first 3 billion years in this landscape's history than the next 570-million-year period, which ended with the arrival of the last ice age. Conclusions made about the northern shield landscape during this time are largely the work of obscure scientists and dreamers. This is because much of the evidence from billions of years ago, the Precambrian bedrock, is still with us today. All of the landscape components that took up residence on top of this rock were scraped away, ploughed under, floated downstream or otherwise sent packing. Soils, fossil plants and animal bones, pollen grains, anything that scientists usually use to reconstruct past landscapes — all gone.

Water, flowing or frozen, was the main agent of destruction. Even before the end of the Proterozoic eon, all of the Precambrian mountains had been reduced to broad, flat rock plains by the erosive effects of water. Flowing, it formed wide-ranging streams and rivers that washed fragments of weathered rock from the uplands to the lowlands and, ultimately, from the lowlands to the sea. Frozen, it formed great continental glaciers, which scraped and gouged and peeled off loose rocks, pushing them sometimes thousands of kilometres away or grinding them to dust.

The Proterozoic eon ended with one of the world's first great ice ages, icing on the cake of a rocky landscape 4 billion years in the making.[19] The meltwaters of these glaciers fed the widespread seas of the early Paleozoic era, which began about 570 million years ago. At the bottom of those seas, trilobites left their armoured skeletons and telltale squiggles in the muck. About 500 million years ago, the planet's first fishes, corals, clams and sponges began adding their skeletons to the growing pile of organic debris on the sea floor. Eventually, the ancient basement of the Canadian Shield became cloaked in a thick mantle of sedimentary rock, particularly limestone, a product of life's early experiments in warm, clear-water seas.

Another ice age engulfed the planet around 435 million years ago. This time, our corner of the Canadian Shield probably did not succumb to the spreading sheets of ice. From the magnetic fields recorded in the rocks, geophysicists have discovered that, back then, the South Pole rested over what is now the Sahara Desert. Fossil evidence of collapsed pingos formed just before this ice age testifies to the presence of permafrost in an area that today is scorching desert. The North Pole was out to sea, over the spot where the Fiji Islands sit today. This puts the equator running smack through what now is the northern shield, an area that may have served as an evolutionary hotbed for the many life forms that blossomed during the last half of the Paleozoic era.

Around this time, the shield rose once again above the waves. Its limestone plains turned green as plants emerged from the sea. The shield's first forests consisted of giant tree ferns and horsetails. These forests rose and spread widely over the virgin landscape. They fell to flooding caused by shifting seas and to forest fires sparked, then as now, by lightning (the oldest fossil record of a forest fire dates back 400 million years[20]). Today, in the Mackenzie Valley, the reconstituted remains of these plants are pumped out of the ground in the form of oil and natural gas.

Around 330 million years ago, the world's climate cooled enough to trigger another major ice age, a global cold snap lasting about a 100 million years. The first needle-leaved coniferous trees evolved during this time, perhaps in response to the cooler climate. The landscape became dominated by the ancestors of today's spruce and pines. Among their branches buzzed metre-long dragonflies and other newly evolved insects. By the end of the Paleozoic era, reptiles reigned over the forest floor.

One of the world's coldest times was followed by one of its warmest times, the Mesozoic era, the golden age of earth's "middle life." Throughout this time, the world was perhaps bluer and greener than it has ever been; there were no permanent ice caps anywhere. Average temperatures 10°C warmer than today prevailed for 150 million years. Conifer trees continued to be the dominant vegetation for most of this era, reaching their fullest expression in a proliferation of species the world may never know again. Instead of black bears and red squirrels, the main denizens of these coniferous forests were the gargantuan reptiles we know as dinosaurs, who dominated the land and sea throughout the Mesozoic. By 140 million years ago, the world's first feathered birds were skittering above their heads, unimaginable cries emitting from their toothy beaks.

By 80 million years ago, the North Pole had wandered to a point about 300 kilometres off the northwest coast of Alaska. This put the Arctic Circle lying on about a 45° angle between Great Bear and Great Slave lakes — not too far from its present position. Like now, this region was a land of midnight sun in summer and long, dark nights in winter. But it was balmy year round. Fossil deposits dug up near Great Bear Lake contain pollen from the bald cypress, various tree ferns and other species that today grow in the subtropical forests of Central America. Fossil alligators, tortoises and boarlike tapirs found on Ellesmere Island point to the kind of animal life that may have inhabited the forests of this region during the late Mesozoic era.[21]

In those days, plant growth must have been vigorous, thanks to temperatures comparable to those of California today and over 20 hours of sunshine in summer. But although the climate was much less severe than now, winter was still a time of pronounced stress. To survive the long period of relative darkness, a time when there was no hope for photosynthesis, plants must have become dormant for many months, as they do now.

Some botanists believe that in adapting to the stress of annual light-dark swings, northern plants were among the first to adopt the habit of growing broad deciduous leaves in summer to maximize solar collection and then shedding them in winter to minimize moisture loss during dormancy. The late Mesozoic was a time of evolutionary blossoming for deciduous trees, including the precursors of today's poplars, willows and birch. It may be that much of this creativity happened north of 60°.

The close of the Mesozoic era was marked by pronounced upheaval on several fronts. Another major round of mountain building began along the continent's western and northern flanks, partly in response to the tremendous load of sediments washed off the Canadian Shield. The coniferous forests were overrun by the newly evolved group of deciduous trees, which have dominated the earth ever since. The dinosaurs went suddenly and mysteriously extinct and mammals rose to ascendancy. And the climate took a sudden dive for the worse.

What caused the brief but pronounced cold spell 65 million years ago has been a topic of scientific controversy for years. It didn't last long enough to bring on an ice age. But it was intense enough to suggest some kind of climatic cataclysm. The most favoured explanation has a giant meteorite crashing into the earth with an impact equalling 10,000 times the power of all the world's nuclear weapons. According to this scenario, a global dust cloud was kicked up by the blast, enshrouding the planet in prolonged darkness and cold. When the dust cleared and things started to warm up again, millions of life forms, including dinosaurs, had been wiped off the face of the earth.[22] Lucky for us, a few rodentlike mammals managed to squeak through this crisis. Their subsequent rise to ascendancy marks one of the main story lines of the next chapter in the earth's history, the Cenozoic era, better known as the age of mammals.

At the opening of this era, a vast temperate forest covered the whole Northern Hemisphere. Palm trees grew in Alaska. Maples grew in Greenland. Sequoias grew in Spitzbergen off the Siberian coast. In Canada's Western Arctic, right up to Banks Island, modern-day spruce and pine grew alongside a host of deciduous trees, such as elm, witch-hazel and hornbeam — species more typical of the country's deep south.

This northern temperate forest hung on for around 50 million years in spite of a series of climatic wobbles, which occurred several times throughout the Cenozoic era. The first major cooling episode arrived 55 million years ago, the beginning of a long-term climatic descent that culminated in the last ice age. By 25 million years ago, sea ice had formed around the coast of Antarctica. On its mountains were glaciers, the first ones formed on the planet for over 180 million years.

The Northern Hemisphere was slower to cool, but by 15 million years ago, glaciers were also forming in the Mackenzie Mountains. Around 10 million years ago, the deciduous element of northern forests declined rapidly, yielding to the cold-adapted conifers. This "proto-boreal" forest remained extensive for several million years, likely ranging as far north as the arctic coast and beyond into the islands. It was home to many kinds of familiar mammals, including primitive beavers, bears and horses, plus not so familiar ones, such as sabre-toothed tigers and their kin.

About 3 million years ago, there were signs that the climate was hardening. Open, treeless plant cover became more widespread — more akin to the dry, grassy steppes of Siberia than today's tundra. Permafrost settled in as a characteristic feature of boreal ecosystems. Sea ice became a semi-permanent fixture over the Arctic Ocean. In the mountains of Greenland and North America, glaciers began to slowly push their creaking lobes of ice forward. The most recent ice age was about to begin.

4

Ice Time: The Pleistocene Glaciation

From a geological perspective, the last ice age was a frozen drop in the bucket of time. Ice ages are relatively rare events, taking up only a tiny fraction of the earth's total history. For instance, before the first continental ice sheet advanced over the northern Canadian Shield 600,000 years ago, this region had remained unglaciated for one quarter of a billion years.

However rare the arrival of an ice age may be in the big picture, this brief chapter in the region's history brought rapid and unprecedented change. The ice age was, and still is, the primary factor controlling its ecological unfolding, from climate and soils to the development and distribution of its present flora and fauna, including *Homo sapiens.*

It is tempting to think of the start of an ice age as a time of intense cold and heavy snowfall. Ironically, 2 million years ago, the winters might well have been warmer and less snowy than they are now. All that was required to get the ice sheets forming and in motion was a slight lowering of average global temperature by about 2 or 3°C. This resulted in summers that were not warm enough to melt the previous winter's snow. Inexorably, each year's snow accumulated, building up huge continental ice sheets even though winters might have been relatively mild. Once gradual cooling began around the beginning of the Pleistocene epoch, about half a million years passed before the ice sheets got rolling over northern Canada.

This was the Nebraska glaciation, named after the place where its evidence is best defined. It lasted for 50,000 to 100,000 years, before summer temperatures warmed up enough to burn off the ice. This cycle happened four times during the Pleistocene, each glaciation being followed by an ice-free "interglacial" period. At times, these periods were warm enough to return coniferous forests to high northern latitudes. On average, temperatures were high enough to bring the planet's ice load down to much less than it is now, a situation that makes some scientists suspect that the next glaciation may be just around the corner.

What triggers the freeze-thaw cycle during an ice age is something of a cosmic mystery. Several theories have been suggested to explain the subtle temperature fluctuations that set the glaciers advancing and retreating. Is our solar system travelling through some kind of galactic dust cloud that occasionally absorbs enough of the sun's energy to initiate glaciation? Does the sun itself vary enough in its energy output — because of sun spots maybe — to pull the trigger? Or is it minor changes in the shape of the earth's orbit around the sun or the tilt of its axis that set the climatic pendulum swinging? It may be a combination of these factors. It may be something else altogether. Whatever the trigger may be, when it went off at the dawn of the Pleistocene, it opened a period of remarkable transformation.

For most of the Mesozoic and Cenozoic eras, a wide, natural causeway joined North America and northwestern Asia. Throughout this time, temperate species of plants and animals migrated freely east and west across this link known as Beringia. The early Pleistocene saw the start of a very different trend: a sudden wave of Eurasian species into North America with relatively few species headed the other way — and why would they? The Asian heartlands were likely colder than North America, as they are today. As the world cooled, Eurasian mammals may have been better adapted for this change than their eastern counterparts, giving them a competitive advantage when spreading into North America. So goes one explanation for the one-way traffic.[23]

Among the migrants were the first deer to set foot in North America, precursors of modern caribou and moose. The first lynx, grizzly-like bears, weasels and wolverines arrived about the same time, along with several new species of hares, squirrels and small mammals. The basic faunal building blocks of today's boreal forest were in place. These animals inhabited a forest dominated by such familiar species as spruce, pine, birch and alder. For a visiting time traveller from twentieth-century shield country, much would have been recognizable. On the other hand, much would not.

A fairy-tale-like array of more exotic species formed a large part of the parade of mammals flowing in from Asia: woolly mammoths, woolly rhinos and woolly oxen with massive curling horns; single-toed horses, wild asses, hyenas and saiga antelope. At the heels of many of these species were two new kinds of sabre-toothed tigers and the scimitar cat. These were the first arrivals of the Pleistocene megafauna, so-called for their relatively grandiose stature. They added a richness of mammal species, both herbivores and predators, unparalleled anywhere in contemporary North America.

Many of these animals were grassland dwellers, having evolved on the central steppes of Asia. As the climate here grew cooler and drier, they found increasingly familiar habitat, for steppes were now encroaching on the conifer forest. During glaciations, these animals lived south of the ice as well as in ice-free northern refuges primarily in Alaska, parts of the Yukon and northwestern Northwest Territories. During interglacial periods, they roamed widely across the continent, including, no doubt, the northern Canadian Shield. Since glaciers stripped this region of most of its fossils, evidence of their presence here is virtually non-existent. But mammoth teeth have been found on several sites near the western margin of the shield. Horn fragments of the saiga antelope have been found east of the Mackenzie Delta. And bones of the Yukon wild ass have been found along the Mackenzie Valley. Until more direct clues are discovered, the best tool for reconstructing a picture of shield country's Pleistocene fauna will be your imagination.

A two-ton glacial erratic lies stranded on the snow-swept shore of Great Slave Lake. During the last advance of the continental ice sheet, it was plucked from a weak spot in the shield bedrock. *(John Poirier)*

The last advance of ice, the Wisconsin glaciation, began about 70,000 years ago. It came from three main fronts: the Greenland ice cap, Hudson Bay — which at that time may not have been a bay at all but a flat, sedimentary plain — and the western mountains. It was not a continuous white tide of ice. Climate is never constant, and over the next 50,000 years the ice sheet advanced and retreated more like an amoeba than a wave. Full-grown forests likely became established during the warmer spells, only to be overrun and crushed by surging lobes of ice as the climate cooled once again.

During one of the Wisconsin's many partial retreats, another influx of mammals crossed from Beringia to North America. Among this bunch were modern-day caribou, moose, foxes, muskoxen, arctic hares, arctic ground squirrels, voles and lemmings — practically all of the species that now characterize northern Canada.

The last ice sheet reached its greatest glory about 20,000 years ago. It covered 15 million square kilometres, about the size of the ice sheet now covering Antarctica. It was more than a thick glaze of ice over the land. The ice sheet had an undulating topography, complete with domelike hills and gently sloping saddles. At its thickest point, over Hudson Bay, the ice sheet may well have been 5 kilometres from top to bottom. Downstream, over the northwestern Canadian Shield, the ice probably averaged around 2 kilometres in depth. Except for a few pupating insects or dormant seeds trapped in caves beneath the ice, this land, like much of the upper Northern Hemisphere, was stone dead.

A little less than 20,000 years ago, a mysterious force started pulling on the planet's climatic pendulum, triggering a gradual swing back into a period of warmth and renewed life.

5

The Emerging Land: Postglacial Flora and Fauna

The warming was slow. After rising imperceptibly, year after year, summer temperatures eventually crossed the critical threshold above which an annual net gain of snow switched to a net loss. The ice began to melt.

It takes a long time to melt a continental ice sheet. At first it just sat there, sagging a bit, getting thinner, washing unfathomable volumes of meltwater into the sea, but showing no signs of retreat. Not until around 18,000 years ago did the ice start shrinking in earnest. Like the advance, this process was spasmodic. There were periods when it slowed to a halt, then readvanced, then backed up, only to readvance again. At the best of times, the pace of glacial retreat was less than 100 metres per year.

Around 12,000 years ago, some great calamity began striking down many of the mammal species that, for almost a million years, had ebbed and flowed south and north in rhythm with the four glaciations of the Pleistocene. In her book *After the Ice Age — The Return of Life to Glaciated North America*, the Canadian botanist E.C. Pielou gives this eulogy to the remarkable disappearance of the Pleistocene megafauna:

> The most striking ecological change marking the end of the Pleistocene epoch in North America was, sad to say, a great loss. In the space of three millennia at most, in the interval from 12,000 to 9,000 [years ago], between thirty-five and forty species of large mammals became extinct. This wave of extinctions is one of the most noteworthy, and

most puzzling, events in ecological history. The reasons for it have been debated for decades, and none of the many explanations put forward is entirely satisfactory.[24]

After dissecting several of these explanations, from overkill by early humans to catastrophic environmental change, she concludes that the reason the megafauna disappeared forever is still an "unsolved puzzle."[25] Far be it for me to add to this debate. What *is* relatively certain is that these animals never had a chance to recolonize the northern Canadian Shield before it was all over. When the last woolly mammoth bit the virgin dust 9,000 years ago, much of this part of the world was still under ice. The rest of it was underwater.

While the megafauna were declining, a megalake was on the rise over the northwestern margin of the shield. During a cold spell beginning around 11,000 years ago, the retreating ice front came to a temporary halt right along this margin. Under the tremendous weight of the ice sheet, the shield rock east of this front was warped perhaps 100 or more metres downwards from its present elevation. Meanwhile, land to the west recently freed from this weight was rebounding upwards. As a result, the huge volumes of meltwater gushing from the glacier became trapped by cliffs of ice on one side and rising land on the other. Glacial Lake McConnell was born.

At its biggest, about 10,000 years ago, this lake was 238,000 square kilometres in area and 1,100 kilometres in length — longer than any freshwater lake in the modern world. It overflowed and joined the basins of the three great lakes of this region: Athabasca, Great Slave and Great Bear. In places, it rose 275 metres above the present lake level. For comparison, that's about a 60-storey building. The region's first lake trout, whitefish and northern pike were moving into a realm where eagles and ravens now soar.

Plugging this whole system was an enormous block of stagnant ice left abandoned by the retreating glacier near Camsell Bend, a bottleneck in the Mackenzie Valley where the river takes a sharp right turn towards the Beaufort Sea. This ice dam forced water from the Liard, Peace and Athabasca rivers to detour into the Great Slave Lake area, then northward to Great Bear Lake. From there, water drained back into the Mackenzie Valley via the Bear River and a now-defunct channel from the northwestern arm of Great Bear. Even after the ice dam disintegrated, Glacial Lake McConnell persisted for several hundred years, blocked later by a massive delta formed at the outlet of the Liard River near Fort Simpson.

As the ice sheet shifted, surged and wasted away, the lake changed its shape and drainage patterns from day to day. Icebergs calved along its eastern shore. Swift, sediment-choked rivers writhed like snakes in front of the ice, as old channels became blocked with glacial debris and new ones carved their way down to the lake. Over its lifespan, the lake switched drainage directions several times. For a while, it emptied south into Lake Agassiz, another proglacial lake covering much of Manitoba, then north into the Mackenzie, then back and forth a few more times until everything finally came apart around 8,700 years ago.

It was a sudden but not surprising death. The land to the west of the lake continued to rise, though not at a uniform rate. The lake's northwestern shore was rising the fastest, eventually choking off its main outlet channels. Something had to give. Tremendous water pressure built up quickly behind the lake's key plug, the Liard River delta. Once water overtopped the delta, the lake's fate was sealed. The soft, gravelly sediments provided little resistance once water started bursting through. What followed was a catastrophic flood, the likes of which today's world has probably never seen. Near Fort Simpson, cleanly incised escarpments 30 metres high bear witness to the power of this event.

When the flood was over, three daughter lakes were left behind — Athabasca, Great Slave and Great Bear — lesser shadows of the once great Glacial Lake McConnell. Scattered inland from their present shores are raised beaches and boulder terraces sorted by ancient waves. As glacial rivers emptied into Glacial Lake McConnell, they dropped their coarser sediments of sand and gravel near the shore. One such river flowed down what is now the main street of Yellowknife, depositing the sands on which much of the town is built. Finer sediments were carried into deeper waters, where they eventually fell to the lake bed, creating laminated pockets of clay and silt that now rest high and dry above the water.[26]

Ancestral Great Slave Lake had the same basic profile that it does today, except that it had three arms instead of two. Its present shape gives it the impression of a swan in flight, the East Arm being the head and neck and the North Arm being one of its extended wings. The other wing was clearly visible 8,700 years ago. The extreme south shore of Great Slave Lake once came within 10 kilometres of the future town site of Fort Smith. What now is the Slave River valley was filled to the brim with water.

This south arm was a very active environment back then. Sedimentation at the mouth of the Slave River was the main land-building process. Fed by erosion off the newly exposed landscape, the delta edged northward at the remarkable rate of about 20 metres a year. Meanwhile, postglacial rebound of the land surface contributed to this process, particularly for the first 2,000 years after Glacial Lake McConnell subsided. During this time, the whole valley literally was lifting out of the water at a rate of almost 2 centimetres a year. Together, these processes transformed a 180-kilometre-long arm of Great Slave Lake into terra firma in just over 8,000 years.[27]

Soon after the death of Glacial Lake McConnell, the last trace of glacial ice disappeared from the northwestern shield. The sun shone down once again on bald Precambrian rock — perhaps for the first time in several hundred million years.

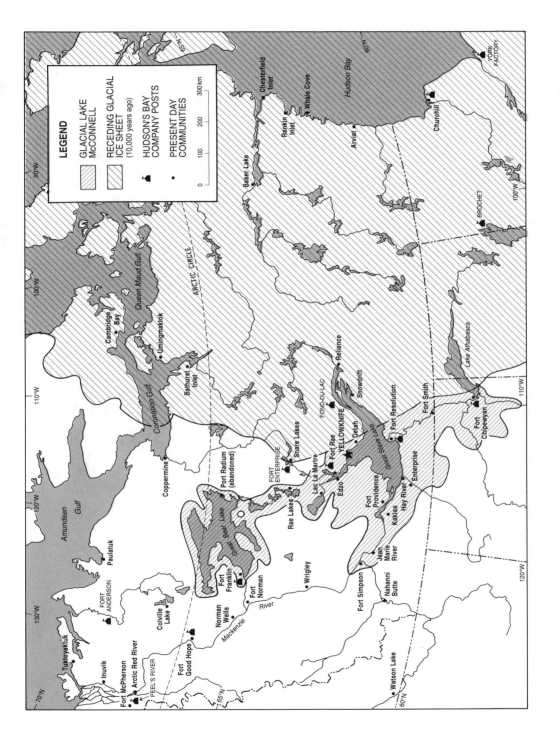

Glacial Lake McConnell. If this oceanic megalake were to rise again, it would inundate over 50 percent of the communities in the Western Arctic, in some places to a depth equivalent to the height of a 60-storey building.

Like cracks on the surface of a cake rising in the oven, many of the random clefts in shield rock have occurred as it rebounds upwards, unburdened by the tremendous weight of glacial ice. *(John Poirier)*

The process of stripping the shield of its cover probably began in earnest during the early days of the Cenozoic era. With the birth of the Rocky Mountains, many large rivers must have flowed right across the shield on their way to Hudson Bay. Like a northern Mississippi, a Liard–Thelon River may have been the main artery for this flow, carrying with it all of the sediments removed from this landscape by innumerable tributaries. As water erosion lightened its load, the entire Canadian Shield began uplifting, in places by as much as 200 metres. In turn, erosion became more intense as streams flowed faster off the higher ground.[28]

In this way, water erosion sliced through generations of soils, past the fossils of creatures long extinct, past the coal and oil formed from ancient tree fern forests, past the limestone mantle laid down by Paleozoic seas. Over millions of years, water carved down to the Precambrian basement rock. The last wave of glaciers finished off this process, laying bare once again the stable, unchanging nucleus of the continent.

Though the glaciers took much while advancing, they also left behind considerable debris in the wake of their retreat. Scattered randomly over the bedrock were corrugated piles of morainal till and patches of clay, sand and gravel dumped under the ice, out of glacial rivers or on the bottom of proglacial lakes. These sites provided a toehold for the return of plants.

The process of recolonization was probably quite rapid, even though the first plants were starting more or less from scratch. This region's emergence from under the ice coincided with a period in history which, since then, has never been warmer. This was the so-called Hypsithermal, the great warmth. It lasted from 10,000 to 6,000 years ago, having its greatest influence in this part of the world about halfway through. During this time, summer temperatures may have been as much as 12°C warmer than they are now. As a result, the growing season was significantly longer, by a month or more, and permafrost was probably non-existent.

Seed sources for colonizing this region were certainly not in short supply. Even while shield country was still entombed in ice, white spruce trees of impressive stature had long since established themselves as far north as the Tuktoyaktuk Peninsula. The great warmth had hastened their migration northward and, in some places, pushed the tree line almost 350 kilometres beyond its present position.[29] By this time, most of the species making up the modern boreal forest were probably well established in non-glaciated or non-flooded parts of the Western Arctic.

The first tree colonizers of this region were likely poplars and jack pines, since they would have done best on the coarse, dry-textured pockets available. As soils developed, spruce, birch and a host of shrubs would have been close behind, bringing with them the distant ancestors of today's animals and birds.

Within a few decades of their establishment, the first forests were burning. Since then, they have probably burned 1,000 times over. The oldest known forest fire in shield country burned forests beside Roundrock Lake about 7,000 years ago. Just what sparked this fire is hard to tell. It might have been lightning, as it had been for millions of years. Or it might have been caused by a new kind of flame that arrived on the scene right around this time, a flame that escaped from a hearth built by human hands.

6

Our Species Arrives:
The Peopling of Shield Country

While the world's climate was taking its final plunge into the Pleistocene ice age almost a million years ago, our ancestors, in the form of *Homo erectus* — the "upright man" — were headed north, out of Africa, into the cold. They carried with them a powerful new knowledge that opened up their world even as it began to freeze over. They had mastered the technology of fire.

By 800,000 years ago, their campfires were lighting up the foothills of the European Alps. Some 100,000 years ago, they were burning in the bamboo forests of northern China. By 50,000 years ago, a newly evolved species, *Homo sapiens*, was roasting mammoth meat over the fire on the dry steppes of eastern Siberia.

Soon after this time our species first entered North America via the Bering land bridge. These were the Palaeoindians, the so-called "mammoth hunters" or "ice edge people," who by 11,500 years ago had spread from Alaska into the American Southwest. Just how they got there remains an archaeological mystery, one pondered long and hard by Chris Hanks, a subarctic archaeologist who has logged more than ten summers on field sites north of 60°. To him, the big question is, "Okay, you've got people in that northern corner of North America. How do you get them south of there?"

Hanks believes that inland passage south through the mountains was blocked by the ice sheet until around 11,000 years ago — 500 years *after* the well-dated presence of humans south of the ice in the American Midwest. Around this time, the Cordilleran and

Laurentide ice sheets parted, creating an ice-free corridor stretching from the Mackenzie Valley to the Peace-Athabasca area of western Alberta. Even if this corridor was open earlier, it must have been a forbidding place to travel. In *After the Ice Age*, E.C. Pielou conjures up the scene faced by would-be immigrants:

> Incessant, strong, bitterly cold winds would have blown off the nearby ice sheets. The land would have consisted of a mixture of bogs, marshes, islands of stagnant ice, and rock-strewn barrens, with dangerous, torrential rivers and icy lakes. Such vegetation as managed to grow would have been sparse. Food, clothing, and firewood would have been impossible to come by. While it was like this the corridor could not have served as a home for humans, and a home rather than a migration route was what was needed.[30]

Not a pleasant trip. As an alternative route, Hanks suggests that the first southbound Palaeoindians may have migrated along ice-free shores west of the Rockies. He admits that, although this is a convenient argument, it is difficult to prove since much of this route was inundated with seawater once the ice sheets melted. Pushing the bounds of possibility yet further, he offers a third option: "It may in fact be that they migrated south from Beringia 30,000 years ago, before there was any ice to worry about. If that's the case, finding evidence of this migration will be very difficult because of the ice that followed them south."

However they got there, the Palaeoindians apparently thrived in their new world south of the ice. In just a few hundred years they spread rapidly across the continental United States, leaving behind their distinctively fluted "Clovis" spearheads from Arizona to Pennsylvania. As the glacial ice retreated, they followed the last of the mammoths and long-horned bison north into Canada.[31]

The oldest known evidence of Palaeoindians in the northwestern Canadian Shield was discovered in the early 1960s by the Acasta River, situated about 130 kilometres southeast of Great Bear Lake. This is the same Acasta River where, 30 years later, the world's oldest known rocks were discovered — a place becoming famous for old things.

At the base of a tall, sandy esker near the shore of the river is a natural amphitheatre that once offered shelter to some of the region's first human residents. This was a popular spot in those days. In an excavated area of only 80 square metres, archaeologists unearthed no fewer than 105 fire-reddened hearths. In some spots, these hearths were superimposed on each other seven layers deep, suggesting recurrent use of this site by several generations of Palaeoindians, or the Shield Archaic people, as they have come to be known by archaeologists.[32]

In some of their hearths, 7,000-year-old charcoal was mixed in liberally with fragments of quartzite, a particularly hard, silica-rich mineral that flakes more easily once roasted. This made it a favourite resource with Stone Age artisans fashioning spearheads.

In the same hearths, charred bones indicated that the Palaeoindians were no longer taking aim at mammoths, ground sloths and other Pleistocene megafauna (they had been extinct for 2,000 years) but at barren-ground caribou, black bear, beaver, hare and eagles — a mix of animals from both sides of the tree line.

Former mammoth hunters were now caribou hunters, an ice-edge people was now a forest-edge people. The Shield Archaic culture had evolved from its Palaeoindian roots to adapt to the opportunities and demands of life on the northern shield. The climate was changing too, for the worse.

The first humans arrived in this region at the tail end of the Hypsithermal — the great warmth — a period when the tree line was well established far north of its present-day position. The beginning of the Sub-Boreal period, 3,500 years ago, was a time of pronounced cooling that forced the tree line into retreat, in some areas, particularly east of Great Slave Lake, to positions far south of today's forest edge. During this period the Shield Archaic people abandoned the northern shield. They probably left because the deteriorating climate pushed the tree line too far south for them to continue commuting to their favourite caribou hunting grounds on the barren lands. For instance, evidence from several sites along the Thelon River shows that they packed up and walked away for the last time soon after the Sub-Boreal period began.

The Shield Archaic people drifted south and east, providing the cultural rootstock for all of today's Algonquin-speaking people, including the Cree and Ojibway. Meanwhile, their former territory was taken over by a different breed of people, the Palaeoeskimos, who probably moved down from the Coronation Gulf area and Victoria Island. Their ancestors, the earliest Inuit, first set foot on the continent around 4,000 years ago, long after the Bering Sea swallowed the land bridge from Asia. Either they made a run for it across the sea ice in winter or, more likely, they arrived in summer in small armadas of skin boats, the original sea kayaks. Their on-board "tool kit" (as archaeologists say) would have included curved, palm-sized flesh scrapers, a design still used today in the form of metal-tipped ulus. They used these scrapers as often on seals and whales as caribou, for their culture made the best of two worlds, the seacoast and the tundra.

With the arrival of the Sub-Boreal cold snap, the Palaeoeskimos took their caribou hunting skills south and adapted themselves to yet another way of life along the northern edge of the boreal forest. Scrapers, spearheads and stone drill bits from their distinctive tool kit are found as far south as Lake Athabasca, many hundreds of kilometres below the present forest border. The cooler climate may have forced them south, with increased sea ice disrupting the migratory patterns of their marine mammal prey. Or they simply may have been taking advantage of an empty territory with plentiful game. Whatever the cause of their original migration, they ended up ranging across the entire length and breadth of shield country for the next 1,000 years.

"The belugas are back! The belugas are back!" This might have been the message that spread gradually through the region, from camp to camp, triggering the exodus of Palaeoeskimos back to the coast. Their cultural signature in the archaeological record ends 2,600 years ago, about the time when a completely new signature begins. Short, wide-stemmed spearheads and narrow whetstones made from red slate are among the tools announcing the arrival of the Athapaskans, blood relatives of the Dene people who now occupy this land.

The original Athapaskans were probably among the last immigrants to make the land journey across Beringia before it washed out 10,000 years ago. Even though by this time migration to southern parts of the continent was virtually unrestricted, the Athapaskans lingered on in Alaska for at least another 5,000 years. Chris Hanks believes that what got them moving to other lands was a sudden catastrophic change in climate, which he likens to a localized nuclear winter.

The detonator was a mountain called Mazama in southwestern Oregon. According to Hanks, "It blew the entire mountain apart. The ash plume was so huge that it decimated the surrounding landscape and chilled things down for a while over much of the northwest." In search of a more temperate climate, the Athapaskans began their migration towards the continent's interior.

Ecologically, the shield country they eventually arrived in 2,500 years ago was pretty much identical to the shield country of today. The tree line had settled down more or less to its present position. The yearly pattern of caribou migrations was as predictable as it is now (which sometimes isn't very). Lake trout and whitefish spawned in the same streams and sheltered bays. Moose waded through the same rivers and shallow lakes. In summer, frequent forest fires marched across the landscape, constantly changing the age and species composition of the forest. During the long winter, the land was blanketed with thick, powdery snow.

More than any other season, winter must have placed its stamp on the early Athapaskan culture. Without a reliable source of fire, the means and materials to make well-insulated clothing and a high daily intake of food calories, the land would have been uninhabitable. The pre-eminence of winter in the lives and culture of early peoples is preserved in the opening lines of a contemporary Dene creation story:

> In the beginning, man did not exist. Then suddenly there was man, it is said. When winter came, man made himself some snowshoes.[33]

We have no evidence that the first Athapaskans in shield country made snowshoes. The region's oldest known pair hangs in the collections of Yellowknife's Prince of Wales

Northern Heritage Centre and date back only to 1929. Yet the tradition must be thousands of years older, or else how did early peoples manoeuvre through the deep snow, how did they pursue moose and caribou and bring meat back to the hearth? As for summer travel, there is absolutely no material evidence that these people made birch-bark, spruce-bark or moosehide canoes. Yet surely they were plying the innumerable lakes and rivers of shield country from the earliest days; otherwise how could people make the long seasonal trips to their favourite hunting grounds on the barren lands or their fish camps on the big lakes and rivers?

Reconstruction of the early Athapaskan way of life relies heavily on such deductions. Being entirely dependent on the cycles and movements of the animals with whom they shared the forest, they were nomadic. Being nomadic, they could not afford to be a material-intensive culture. The most useful articles of their culture were light and easily constructed from natural materials at hand. If something was not immediately useful it was tossed aside, only to rot away or disappear into the muskeg. Hence almost everything made out of wood, animal skin, feather, bark or bone — probably the basic raw materials of their cultural tool kit — is gone.

All that is left to us of the earliest Athapaskans are a few odd bits and pieces of stone tools, reflecting only a glimmer of their story. However meagre this evidence may be, one message from the distant past comes through clearly. The climate, biology and physical characteristics of the northern shield landscape placed precise demands on its early inhabitants, demands that allowed little deviation or cultural experimentation from one generation to the next. For over 20 centuries, the archaeological record shows little significant change in the type of tools used by natives of this region. This suggests that, once the rules of survival in this environment were learned, there was no point in trying to bend them.[34]

By meditating on the few remaining Athapaskan artefacts and extrapolating back-wards through time from the region's first historical records, archaeologists have pieced together the likely cycle of life for taiga shield dwellers over the past two millennia, a cycle still echoed in most Dene communities today.[35]

The most sociable stage of the cycle was summer. Rich open-water fishing sites provided a regular and abundant source of food, permitting up to several hundred people to gather together in one spot. This was a luxury not possible in other seasons, since big game hunting could support only smaller, more mobile groups. Summer was therefore a time of intense social activity. Most rites and festivals were celebrated at this time of year, usually accompanied by the beating of skin drums.

With the first signs of autumn, people dispersed from the fish camps to hunt caribou and moose. In preparation for winter, they made good use of the animals' abundant fat

Bow hunting caribou 2,500 winters ago. The original Athapaskan inhabitants of shield country had a profound life-sustaining knowledge of this land, which is reflected today in many of their Dogrib and Chipewyan descendents.

and well-furred hides, essential gifts of the land available only in this season. To intercept the caribou as they moved south, many groups ventured out beyond the tree line and set up temporary camps on the barren lands. Movement back into the forest usually happened well before freeze-up, since birch-bark canoes do not mix well with floating pans of sharp, thin ice.

In their winter camps, early natives relied on stores of dried meat and any game that could be taken nearby. With a good fall hunt behind them and plentiful game passing through, they could last out the winter in one camp. But usually they had to move their hide shelters several times to make up for local depletions of meat. When big game was especially scarce, they switched their hunting effort to snowshoe hare, grouse and ptarmigan. The problem of relying on smaller animals is that none of them have much body fat, the resource so crucial for winter survival. To make matters worse, their populations are subject to sudden and severe declines. The grimmest months were January and February, when intense cold and prolonged darkness took its toll on the hunter's success rate, whether pursuing big game or small. A common visitor during this part of winter was starvation.

The return of ducks and geese in May helped dispel the last hunger pangs of winter and carry people through breakup, another period of relative immobility. Birch bark is at its supple best in spring, the time of year when new canoes took shape along the shore. Once the water was clear of ice, the boats were launched. Besides being proficient canoeists, the Athapaskans were also great walkers. This time of year, some groups would strike off overland with their tumpline packs, following well-worn trails that crisscrossed the shield. As people converged from all directions at the summer fish camps, one yearly cycle closed, another began.

7

Changing Peoples, Changing Land: Historic Times

With one sweep of his royal pen, King Charles II of England declared the northern third of North America open for business, ushering in an era of widespread ecological and cultural change unparalleled since the arrival of the first Palaeoindians. The date was May 2, 1670, and the business was furs — "furs enough to make men forget Cathay."[36] On this day, the king set his signature on the bottom of a royal charter making Prince Rupert and his band of gentlemen adventurers "true and absolute lords and proprietors" of all the land drained by rivers flowing into Hudson Bay. For the next 200 years, Rupert's Land and much of the adjacent North-Western Territory were, on paper at least, the private fur-trading preserves of the Hudson's Bay Company.

With the onset of the fur trade era, all subarctic mammals bigger than a vole or a shrew were now fair game for an economic market thousands of kilometres away. For the original peoples of this land, a centuries-old way of life based on nomadism and subsistence was to be profoundly transformed by a grand commercial venture based on frontierism, resource exploitation and the fashion whims of European high society.

For over 100 years "The Great Company" limited the spread of its trading posts to eastern Canada, mostly hugging the shores of Hudson and James bays. To capitalize on furs from outlying areas, fur traders relied on Cree and Chipewyan middlemen to provide a link to more remote tribes such as the Dogribs and Yellowknives of the northwestern Canadian Shield.

The first company man did not reach shield country until 1771. His name was Samuel Hearne. His mission was not motivated by furs but by the shiny red knives brought back to him at Fort Prince of Wales (now Churchill) by his Chipewyan envoys. The knives had been fashioned by the Yellowknife tribe — so named by Hearne — from native copper obtained from the Inuit. His quest was for the source of this copper. He supposed there to be some "fabulous mine" along the arctic coast and, with luck, a shortcut through the Northwest Passage. Finding neither, he returned despondent to his home fort, but not without first doing some skeletal mapping of a territory hitherto untouched by European hands.

Those rudimentary maps and the increasing rivalry between the Hudson's Bay Company and the upstart North West Company propelled the Western Arctic to the forefront of the northern fur trade. By 1786, the light from candles made in Wales was flickering on the southwestern corner of shield country. In the spring of that year the North West Company trader Peter Pond and his crew had completed construction of Fort Resolution, situated where the Slave River empties into Great Slave Lake. That summer, during his exploration of the lake's north shore, he built a cabin just east of Wool Bay, 20 kilometres southeast of present-day Yellowknife. Though never growing to more than a half dozen humble cabins, the site became an important outpost camp for supplying fish and game to larger posts and, accordingly, was dubbed Fort Providence. Two years later Roderick McKenzie, one of the masterminds behind the North West Company, built the trading post of Fort Chipewyan on Lake Athabasca.

In 1789, Alexander Mackenzie stopped at Fort Providence on his way down the river that today bears his name. With him was another Nor'wester, Laurent Le Roux, whose job it was to stay on and transform local native people into full-fledged members of the trading economy. In his birch-bark canoe were muzzle loaders, gunpowder, blankets, knives, axes, needles, kettles, flour and beads, plus a host of other tools and trinkets that he could trade for fish, ducks, hare, moose, caribou and, of course, pelts, the most coveted of all being the beaver.

So valuable was the beaver pelt that it became the standard currency of trade for over 100 years. Furs, meat, dry goods — virtually anything that changed hands in the bush — was assigned a price measured in MBs, "made beavers," one of which equalled the value of a beaver pelt in prime condition: silky, smooth and unblemished. In those days, a person could lay ten snowshoe hares, four ducks or four moose tongues on the trading post counter and walk away with one MB worth of provisions. A decent knife cost at least two MBs, a down vest four, a blanket eight. Before long, trading companies began minting their own coins. Their denominations were measured in MBs. Although such coins were phased out at the beginning of the twentieth century, this long-standing tribute to the beaver's worth is kept alive on the back face of every modern-day Canadian five cent coin, the nickel.[37]

Throughout the first half of the eighteenth century, beavers were mined from the northern forests. The myth of superabundant fur resources, the lust for short-term profits and fierce competition among companies put long-term conservation measures far from the fur traders' minds, often until it was too late. By 1820, beavers had nearly disappeared from around Lake Athabasca. Other furbearers and large game were also in decline. In his report for that year, Hudson's Bay Governor Simpson was not optimistic about the state of wildlife resources:

> The valuable Fur bearing animals are not numerous in this District. . . . Beaver . . . are rarely to be found within a considerable distance. The large Animals, Buffalo and Deer are also scarce in the neighbourhood of the Fort so that little dependence can be placed on them for maintenance.[38]

One of the forces pushing the fur trade north and west was the need to give over-trapped areas a rest and move into virgin territory, where furbearers and game were still plentiful. Another force was the unbridled monopoly enjoyed by the Hudson's Bay Company following the 1821 takeover of its arch rival, the North West Company. The "Bay Boys" immediately set about consolidating a hold over their new territory. The letters HBC quivered on flags raised over all the existing forts north of 60°. New forts were built. In front of each spread a wave of further beaver declines.

By 1841, company managers were forced to introduce new conservation measures, hoping that beaver populations would recover. They tried to persuade trappers to lay off beavers for a while by offering a premium on smaller furs, such as muskrat and marten. This scheme proved to be a case of too little too late. Beavers remained in short supply. In some areas they were trapped out completely. It took a revolution overseas to reverse this trend.

The revolution began without warning on an evening in 1843 when a highly respected individual from the royal court of England appeared in public wearing none other than a *silk* hat. Felt hats made from the dense inner fur of beaver hides fell quickly out of fashion and never regained their former popularity. Attempts to market the beaver hat in other forms failed.

Although the beaver was out of vogue, the fur trade continued to expand, intensifying its efforts to exploit more fashionable species, such as silver fox and marten. By the mid-1870s, the Hudson's Bay Company controlled a dozen forts in the Mackenzie–Great Slave region, including several that served as collection points for furs taken from the taiga shield and beyond from the barren lands: Fort Franklin, Fort Smith, Fort Resolution and Fort Rae.

Arriving in an armada of birch-bark canoes, Dene trappers and their families deliver huge bales of fur to the trading post at Fort Resolution in 1901. *(Mathers/NWT Archives)*

To operate year round, these forts relied almost exclusively on country food available at hand. Sugar, flour and tea were among the only imported staples. By necessity, all forts were situated in the neighbourhood of a good fishery. A secure source of whitefish, trout and pike was especially important when big game became locally scarce or, after around 1850, there were hungry sled dogs in camp. Most of the thousands of kilograms of meat, hides and fish needed were provided by the Dene. Ironically, they were the ones who suffered most when food became scarce.

At first, the Dene had no need for the fur traders. The Dogribs, Yellowknives and Chipewyans had sustained themselves independently on this land for thousands of years. On the other hand, the fur traders needed them for their economic and physical survival. To get natives to work for them, the fur traders had to persuade the Dene to change their ancient ways of life.

As nomadic hunters, the native people of this region had followed game across the land, taking only what they needed for food and clothing. As servants of the fort, they

An abundant local supply of fish was vital to the winter survival of fort inhabitants. For instance, soon after Fort Chipewyan was established in 1788, the annual take grew to 50,000 whitefish (shown here), trout, pike, perch and goldeye, with 3.5 kilograms allotted to each man per day. *(Busse /NWT Archives)*

were encouraged to harvest far beyond their own needs, giving wildlife little chance to recover. Before the fur trade, they traditionally travelled to the barrens in late summer and fall to intercept the caribou and then follow them south into the trees for the winter. As the fur trade took hold of their lives, many natives were encouraged to reverse this cycle for the good of the company. In pursuit of fox pelts, many moved onto the barrens in winter and headed for the woods in summer — not to pursue caribou but to be at the trading post dock when the supply boats arrived.

After decades of trapping and trading, natives had become "expert beaver hunters," as Simpson called them. But many had forgotten how to provide for all the needs of their families as they could in the days before kettles, axes and Hudson's Bay blankets. During lean times, with game depleted around the fort and poor hunting success out on the barrens, many had no choice but to seek handouts from their supposed benefactors. Within the shadow of the fort they sometimes starved.

From his leather-backed chair in Fort Chipewyan, Factor Roderick McFarlane wrote in 1880 of the desperation faced by the native people of Athabasca and other northern regions:

> . . . owing to the scarcity of food animals, and the comparative failure of the fisheries, numbers of Indians will doubtless suffer many privations betwixt [now] and Spring. We have already expended a lot of fish and potatoes on them; in short, but for the assistance thus annually rendered to starving Indians, throughout the North, many of them would assuredly perish. Even as it is, now and again, cases occur beyond our reach, which from a scarcity of food, result in death! It strikes me very forcibly that something must be done and that speedily to help these poor people.[39]

The fur trade introduced deep frays into shield country's age-old social and cultural fabric. It also changed the face of the land. Besides creating local depletions in furbearers, big game and fish, the influx of explorers and settlers during the late 1700s and 1800s had a major influence on the frequency and extent of forest fires.

Long before Europeans arrived in the region, native people had become skilled in the art of "pyrotechnology," the practice of applying fire to the land for a specific purpose. They often used fire to create open areas favoured by certain species of game such as moose, bear and, farther south, bison and mule deer. Fire control was promoted by burning on windless days in the spring when snow still lay on the ground. When the fur trade got going, some natives used this skill to maintain rich wetland habitats where beavers, mink and many other furbearers thrive. They also used fire to clear the way for overland travel, to create dry firewood and, ironically, to reduce the risks of huge unexpected fires, which could disrupt their traditional movement patterns. Fire was even used as a divining tool to improve hunting success.

> Near Fort Resolution, caribou hunters are known to have built fires where trails forked — one fire close to one trail and a second near the other. After the fires had burnt down, the hunters would choose the lucky trail by noting the pile of burnt remains which best resembled caribou tracks.[40]

To the Europeans, fire was largely destructive and wasteful, a potential threat to the storehouse of furs and food that the forest represented to them. Inevitably, some of the native-set fires got out of hand. Many people attributed all fires to them. The French missionary Emile Petitot, who toured the region in the 1860s, left us a classic example of this kind of thinking. He concluded that every charred forest he came across was the result of "the savage's carelessness." After gloomily observing a burn near Great Bear Lake, he commented that the Dene were "insane" to destroy their land in such a way.[41]

In reality, runaway campfires were probably the main source of human-caused forest fires. When ripe for a burn, the forest made no distinction between sparks from a Dene or European campfire. Expansion of the fur trade and the corresponding increase in mobility and human traffic through the region meant more campfires. And more campfires meant more forest fires, especially along the well-travelled rivers, lakeshores and cross-country trails.

With the objective eye of a British explorer, John Franklin, en route through shield country to the Coppermine River, described the aftermath of a campfire that got away on him. On the banks of some as yet unnamed river, his entourage lit a fire to signal their presence to their "Dog-ribbed" guides.

> A fire was made on the south side of the river to inform the chief of our arrival, which spreading before a strong wind, caught the whole wood, and we were completely enveloped in a cloud for the three following days.[42]

Besides used for signalling, cooking or warmth, fires were often built for bugs. Before the invention of special nettings and pungent repellents, a smoky "smudge" fire was the only means of relief when the mosquitoes and black flies got thick — besides gale-force winds. The early geologist Charles Camsell provides a firsthand account of the northern traveller's occasional dependence on fire to keep sane:

> They rose up in clouds with every step I took. I had no protection from these pests . . . and from time to time as I got tired I also became almost panicky. When I felt myself beginning to run I immediately pulled up and made a small fire so that I could get some relief in the smoke. I could easily imagine a man going off his head if he should have to endure such torture for any length of time.[43]

Under such conditions, you have to wonder if many travellers had the presence of mind to douse the flames before pushing on.

The sparks really began to fly around the turn of the century. Some of them sprang from the smokestacks of the Hudson's Bay Company's new fleet of steam-powered paddle wheelers, launched in the 1880s. A report written several years later took a hindsight look at the "distressing wrack and ruin" caused by steamers chugging up and down northern waterways.

> As soon as the boat has loaded up . . . incandescent fragments of cinders are vomited out of the funnels into the woods with disastrous consequences. Additional steamers are

continually employed on these waterways, all of which require large quantities of wood for their power. A new steamer . . . consumes no less than 2 cords of wood per hour.[44]

Often piled metres high on their bows was wood that had been "ruthlessly hewed down" from towering stands of spruce that once bordered much of the Slave River valley and the shoreline of Great Slave Lake. In addition to their direct impacts along water routes, steamers brought a wave of environmental change to much of the region via their passengers.

Canada was moving west and north by this time, and the official policy of the day was to remove any obstruction to the flow of population into these lands. Steamers did much to increase the flow north, turning a former trickle into a flood. On board these Hudson's Bay Company boats were the first independent fur traders and trappers, the very people who would eventually break the company's monopoly over northern furs.

Close behind the southern traders and trappers came others in search of adventure and riches: tourists, sportsmen, surveyors, miners and prospectors. The floodgate blew wide open in 1898 during the peak of the Klondike gold rush. The winter before, a government publication titled *The Klondike Official Guide* had tipped off gold seekers that they could reach the Yukon's Klondike River via Edmonton and the Northwest Territories, through the "Canadian back door."[45]

In droves, they clambered aboard the northbound steamers. Many found their own way downstream by canoe, on barges or on makeshift rafts. Although many turned back or drowned, by the end of that year almost 900 prospectors had reached Fort Smith. The more intrepid among them ventured on down the Mackenzie River, then up the Liard, the Keele or the Peel and onwards to the Yukon gold fields. Many stayed behind, lured by the promise of furs and rumours of gold closer to home.

When the new century opened, furs were still in great demand. New posts continued to spring up throughout the forested north and trading and trapping intensified. Thanks to the Klondike gold rush, the prospecting business was also flourishing. By this time, oil had already been discovered in the Mackenzie Valley and cobalt and silver on Great Bear Lake. The monumental Pine Point lead-zinc deposit was already drawing much attention from prospectors during this period. Numerous mineral claims had been staked along the north shore of Great Slave, but the gold discovery that put Yellowknife on the map was still over 30 years away.

Nowadays, a prospector can turn to helicopters, satellite images and computer models to get an understanding of the land and read its clues. In the old days, this knowledge was sometimes gained with the strike of a match. To many prospectors, the forest was simply an inconvenient covering that prevented their knowing what kind of rock was underfoot. Commenting on a northern prospector's work in 1898, historian

One of hundreds of prospectors' camps that sprang up throughout the region in the wake of early rumours of gold. *(Busse/NWT Archives)*

Arthur Lower observed that "prospecting for this stuff (gold-bearing quartz) means hunting veins through the rock with pick and dynamite, after having first burned down the forest to let the surface of the rock be seen."[46]

The true impact of increased human activities on the boreal landscape around this time is impossible to gauge. While some desperate prospectors may have torched the forest or overzealous trappers sent too many beavers to oblivion, natural forces such as lightning, disease and population cycles were still operating as always. Some mysterious mix of natural and human-caused factors made this period of history a time of pronounced environmental change. The most distressing trend was the decline in caribou.

Before the decline, thousands of barren-ground caribou were said to pass through the Fort Rae area like clockwork, appearing regularly around All Saints day in early November.

> They were often killed from the buildings, and throughout the winter might be found near the post. In 1877 an unbroken line of caribou crossed the frozen lake near the fort; they were fourteen days in passing, and in such a mass that, in the words of an eyewitness, "daylight could not be seen" through the column.[47]

The first reports of caribou declines around the fort date from the 1880s, about the time when trading and trapping in the region was reaching a fever pitch. Stories of local game depletions and starving natives continued into the twentieth century. In the winter of 1910, the caribou did not show up at all, according to Corporal Arthur Mellor of the Royal Northwest Mounted Police:

> Things were in a most lamentable state at Fort Rae; the Indians were practically all starving, owing to the entire absence of caribou. Father Toure, the priest there, informed me that this is the only time the deer have failed to arrive, during his 42 year stay at the place.[48]

Was the decline due to disruptions of their winter range by widespread fires? This was the conclusion reached by Warburton Pike, who, on a hunting trip near Great Slave Lake in 1889, was struck by the absence of caribou.

> . . . they [now] keep a more easterly route. . . . This is in great measure accounted for by the fact that great stretches of the country have been burnt, and so rendered incapable of growing the lichen so dearly beloved by these animals. . . . Within the last decade, the southern shore of Great Slave Lake has been burnt and one of the best ranges totally destroyed.[49]

On the other hand, was the decline caused by a radical shift in their migration patterns triggered perhaps by a change in the weather? When Charles II set the fur trade in motion in the late 1600s, the Northern Hemisphere was experiencing some of the coldest annual temperatures since the Sub-Boreal period 1,000 years earlier. Lasting until around 1850, this was the Neo-Boreal period, a time during which caribou may have penetrated deeper into the forest for winter protection. Increased hunting pressures brought on by the proliferation of people and posts in the region no doubt contributed to their decline. By the 1920s, great numbers of caribou were being slaughtered for food and for sport by hunters armed with high-powered rifles. This added pressure drove some people to the conclusion that caribou were in imminent danger of becoming extinct.

More powerful and accurate weapons introduced during the first quarter of this century were a boon to native people in meeting their domestic needs for dwindling game. However, these improvements in hunting technology may have further contributed to wildlife depletions created by the commercial demands of trading posts. *(Busse /NWT Archives)*

Without protection it would only be a matter of time till the caribou suffered the same fate as the American bison that once was as plentiful on the western plains as the caribou is on northern ones.[50]

Few if any records were kept that could shed light on the severity and rate of caribou declines during the late nineteenth and early twentieth centuries. On the other hand, for muskoxen killed in the barren lands and traded at Hudson's Bay posts throughout the

Mackenzie District detailed records do exist. They tell a story of a sudden and near disastrous decline in the mainland population of muskoxen.

Much as beaver fell temporarily out of vogue in 1843 with the British introduction of silk hats, so muskoxen fell decidedly *into* vogue in 1860, when their hides, instead of bison, became popular as sleigh or carriage robes in Europe. The demand for muskox hides soared. During the late 1880s — the peak of the "muskox boom" — their hides were worth almost $25, a time when a top-quality beaver fetched only $5 and a marten less than $2. Muskox was clearly the preferred item for trade in those days.

Spurred on by high prices, Chipewyan, Dogrib and Yellowknife hunters ventured to the tree line and beyond to hunt muskoxen in great numbers. They usually set out in winter, often bringing back all that their sled dogs could drag. As stocks of muskoxen declined, they had to travel progressively farther out onto the tundra each year. The hides were piled high in posts across the district, with over 8,200 collected between 1861 and 1898. Of these, a whopping 5,073, or 62 percent, were collected at Fort Rae on the North Arm of Great Slave Lake.

By the turn of the century, the flood of muskoxen hides collected in Mackenzie District trading posts had tapered off to a trickle. The barren lands within their sphere of influence were made more barren by the virtual extermination of muskoxen. This was the biological "brink" that geographer William Barr describes in his book *Back from the Brink: The Road to Muskox Conservation in the Northwest Territories*. After sifting through voluminous records and reports from this dark chapter in the history of northern trading, he concludes that overhunting was critically important in causing this decline.

> All the evidence tends to suggest that the abnormal stress on the population represented by the commercial trade in muskox hides, in combination with normal losses due to predation, severe environmental stress and harvesting by indigenous people for their own consumption, was sufficient to bring the mainland population of muskoxen to a point of near extinction by 1916.[51]

The decline in caribou and muskox populations contributed to by overhunting was a real and widespread phenomenon, one that perhaps came none too soon. In the opening decades of the twentieth century, the Canadian government and fledgling conservation groups were still reeling from the near annihilation of plains bison and were not interested in a repeat performance with other big game species. North America had witnessed several human-caused extinctions of birds in the past 100 years and many northern breeders such as golden plovers, trumpeter swans and whooping cranes were already on the endangered list. As furs became scarcer through overharvesting, their dollar value increased, which prompted, in turn, yet more intense trapping — an ecologically vicious cycle.

In spite of higher pelt prices, it was a "lose-lose" situation for everyone. The traders suffered diminishing fur returns. The trappers had to go farther afield to bring home the fur. The hunters had to spend longer periods in the bush searching out their quarry. Many natives were feeling crowded off their traditional territories by southern trappers as competition for fur and game resources heated up. The political climate was ripe for putting a damper on the exploitation of wildlife.

During the first three decades of this century, trial and error played a major role in the federal government's attempts to develop a management regime for northern wildlife. But throughout this time, its objectives were clear.

> We have under contemplation regulations amending the Game Laws which will have for its object not only the conservation of the wild life of the great Northland but also the conservation of the Indian himself.[52]

Turn-of-the-century sportsmen camped near Fort Smith revel in their take of over 250 ducks. Such extravagant exploitation of wildlife eventually moved the federal government to enact new wildlife conservation measures. *(Yellowknife Museum Society/NWT Archives)*

In 1906, the Northwest Game Act was enacted to ease trapping pressures on furbearers and arrest the decline in big game. For the first time in the North's history, closed seasons were introduced on all harvested species. Though well intentioned, the act was almost impossible to enforce given the vast area and the mere handful of Northwest Mounted Police assigned to the job. Even if someone was caught red-handed with illegal booty, it was difficult to lay charges since blanket exemptions in the act allowed just about anybody to hunt anything anytime "for purposes of food."

Recognizing the weaknesses of the original Game Act and the worsening plight of native trappers, the government gave the legislation substantial teeth in 1917, providing that no person except a "native-born Indian, Eskimo or half-breed who is a bona fide resident of the Northwest Territories . . . shall engage in hunting, trapping, or trafficking in game, without first securing a license to do so." Licensing was the most important feature of the new, improved act, since now it was possible to control the activities of foreign trappers, who, according to one conservationist of the day, were wreaking "great destruction" on the North's wildlife resources.[53]

That same year, the Migratory Birds Convention Act was passed to help protect ducks, geese and other game birds from overhunting. Most other birds were declared off limits completely.

Together, these two acts represented a major step forward in the conservation of northern wildlife. As for protecting native interests, they were, at best, a mixed blessing. Although the Dene didn't need licences and the intensity of trapping was now controlled — on paper at least — the new regulations gave outsiders clear rights and privileges to trap in certain areas. This was an affront to some Dene, who traditionally were free to use the entire country, which long ago had been divided into family trapping areas. In her *History of Wildlife Management in the Northwest Territories*, Jonquil Graves points to other regulatory conundrums facing the Dene.

> Dividing the hunting year into open and closed seasons must have been a mystery to a society which hunted only for food and clothing. Seasonal restrictions went completely against the grain of the Dene way of life. Spring caribou hunts on the barrens were as old as the Dene themselves but the new laws declared them illegal. Killing buffalo and muskox when one was lucky enough to encounter them was considered a great coup and a cause for feasting and celebration, but it was now illegal. Spring hunting of ducks and geese was one of the few ways to obtain fresh meat after a hard winter but the Migratory Birds Convention Act said it could not be done.[54]

To help discourage southern trappers and traders from getting out of hand, the government decided to jack up their licence fees. As of January 1923, the fee for non-resident trappers from Britain was raised from $25 to $75. For all other non-residents, the fee rose from $50 to $150. Trading fees went up similarly, and a "resident" was redefined as someone with not two but four northern winters behind them.

As extra protection for aboriginal people, several large game preserves were established in September 1923, providing "exclusive hunting and trapping grounds for all native Indians, Eskimos and half-breeds." Two of these preserves, the Yellowknife and the Slave, brought immediate benefits to the Dene of shield country. But in the vicinity of every trading post and fort, local decimations of wildlife continued unabated. The situation worsened to the point that the government seriously considered establishing an 80-kilometre-radius game preserve around each one.

Meanwhile, the fur trade proceeded with fresh postwar gusto. The new regulations, fee hikes and wildlife reserves caused hardly a ripple in the northbound flood of trappers and traders. Recent improvements in transportation made coming north a lot easier. New steamers were chugging across northern waters, including the S.S. *Hope*, launched in 1915, which cruised Great Slave Lake and the lower Slave River. The first of countless aircraft crossed the 60th parallel in March 1921 — a pair of World War I Junker biplanes. They belonged to the Imperial Oil Company and were headed for a promising oil discovery near Fort Norman on the Mackenzie. By 1925, the railroad had been pushed well into northern Alberta, much to the relief of those who previously had travelled north from Edmonton by stagecoach.

This was boom time north of 60°, as the record clearly shows. It was fuelled mostly by the soaring price of furs. During the 1920s, muskrat and white fox pelts were 20 times their turn-of-the-century price. In 1929, a white fox fetched $54.15, a silver fox $104.42. That was good money in a time when the mean annual wage down south was around $1,000. Yearly incomes supported by the fur trade were often ten times that amount.

In 1920 there were 110 trading stores in the Northwest Territories, most of them south of the trees. By 1927 that number had almost doubled. In Fort Rae alone, shield country's main trading centre, there were 41 new trading licences issued in 1926. During that same period, the number of white trappers in the region rose from 140 to 500.

The people responsible for enforcing game laws began to worry about an ecological backlash. In a 1923 report to his boss, a Northwest Mounted Police Officer puts his concerns in no uncertain terms:

> White men are infringing on the trapping ground of both Indians and Eskimo and will without doubt, in time, clean out this district of all fur . . . I would therefore urgently impress upon you the necessity of putting a stop to this influx of White Men at once.[55]

The crash came in 1928. Fur returns for mink and muskrat hit rock bottom. A three-year complete closure on beaver was declared. Many trappers and traders from the South headed home, leaving behind native people with little means of securing a livelihood. Some had trapped all their lives and now were unable to fully live off the land as their ancestors had. To make matters worse, rabbits, an important item in the native diet, were at the low end of their cycle and caribou were scarcer than ever.

The culmination came the same year, when the Dene population was hit hard by an epidemic of influenza imported from Europe. By the time the epidemic had run its course, over 300 natives were dead. Though the disease took its greatest toll in the Mackenzie Valley region, it did not bypass the Dene of shield country. At the far end of Great Slave Lake's East Arm, a Chipewyan community near Fort Reliance was almost entirely wiped out.

The 1930s brought economic decline to the northern fur trade, which was perhaps accompanied by ecological recovery for species hitherto intensively harvested. Many Dene, such as this man at Fort Rae (right), continued their trapping way of life — as many do today — in spite of failing fur prices. *(1939, Finnie/NWT Archives)*

The flu epidemic and a greatly weakened fur economy hastened the movement of Dene off the land and into the region's permanent settlements. Though many people living in the region were in a weakened and impoverished state, this was a period of much-needed grace for wildlife populations, many of which were on the rebound now that trapping and hunting pressures were at an all-time low.

The Depression of the early 1930s cast its shadow northward in the form of yet deeper plunges in the price of furs. But by 1933, prospects were looking up in another business. In the unfolding drama of northern development, fur trading was quickly moving into the shadows, while mining began to steal centre stage.

Even as the Depression kept its stranglehold on much of the nation, the region's first mine was opening up on the shore of Great Bear Lake's McTavish Arm. By the summer of 1933, Eldorado Mines was making regular shipments of uranium and silver concentrate to markets in southern Canada and the United States.

While production at Eldorado Mines was entering full swing, geological reconnaissance and mapping work by the Geological Survey of Canada identified some interesting mineral prospects in the neighbourhood of Yellowknife Bay. The flurry of prospecting

Eldorado Mine at Port Radium on the east shore of Great Bear Lake, 1946. Uranium ore destined largely for U.S. military markets was shipped from this mine until the early 1960s. *(Busse/NWT Archives)*

Refined uranium from Eldorado Mine was used in the Manhattan Project during World War II, which led to the detonation of the world's first atomic bomb. This "accomplishment" was celebrated by this sign, which hung proudly in the mine manager's office. *(Busse/NWT Archives)*

that followed climaxed in September 1935, when Bill Jolliffe and his party of greenhorn geologists discovered the gold that put Yellowknife on the map.

One good gold discovery deserves another. Prospectors swept through shield country, erecting a virtual forest of four-sided claim stakes on some of the more promising patches of shield rock. Many claims even extended well out along the bottom of lakes. As another floodgate of gold seekers opened, Jolliffe remarked that "probably there were more prospectors in this area in 1938 than any other part of Canada."[56]

As they spread across the land in growing numbers, so did forest fires — the worst on record. Many of the region's contemporary forests, particularly around Yellowknife, took root in the wake of fires dating back to the 1930s. What exactly caused this rash of bad fires? Maybe it was intentional torching of the forests, a quick and dirty prospecting technique that laid the land bare for closer inspection. Maybe it was the prospectors' many campfires, left glowing as they raced down what they hoped was the "yellow-brick" road.

One theory points to the region's first airplane pilots as the chief cause of these fires. The increased mobility they brought no doubt spread the extent of human-caused fire throughout the region. To early fire rangers, this only made sense. What didn't make sense were the inexplicable blazes that kept popping up beneath regular flight paths, for instance, along the well-travelled route between Edmonton and the Eldorado Mine on Great Bear Lake.

After puzzling over several years' worth of northern fire patterns, a forester from Alberta became convinced that many of the worst and inaccessible fires were due to the habit of aircraft pilots chucking cigar butts out the window. Subsequent tests confirmed his theory by demonstrating that a cigar butt could carry on glowing — and hence ignite a forest — after free falling 2,200 metres through the air.

Around Yellowknife, much of the forest that did not burn in the '30s was devastated in the '40s, especially during the summers of 1942-44. These were also hot times for Fort Smith. During this period, residents witnessed forest fires in all directions, including a series of intense burns due north of town that blackened a swath of shield country 20 kilometres wide and 100 kilometres long.

While visiting the region in 1942, a dignitary from Ottawa reported that "fires galore" had been raging all summer.

> Consequently, a thick pall of smoke, as dense as a bad fog, has hung over the whole country between Edmonton and Great Slave Lake week after week, at a time when air

A small, 1930s bi-plane drops fire-retardant on a fire burning somewhere in the taiga forest of northern Alberta. Although planes were a welcome addition to fire-fighting technology, ironically it was cigar-chucking airplane pilots who may have started many fires beneath the region's well-travelled flight paths. *(Edmonton Air Museum Committee /NWT Archives)*

and river transport is usually very active. For long periods, every aeroplane in the place has been grounded, and small craft like scows have been tied up.[57]

It got so bad at times that airports as far south as Edmonton occasionally shut down due to sky-filling smoke that wafted down from the Northwest Territories.

In some cases, trees spared by the flames were felled by the axe. According to one Yellowknife old-timer, "the amount of wood cut in those early days was astonishing!" In the 1937 frontier classic *North Again for Gold*, Edgar Laytha remarked on the acute shortages of local timber:

> . . . the gold-diggers had shorn the rocks of all timber the moment they arrived. . . . The timber ran out and there was nothing to be done; anyone who still had any, needed it to put a roof over his own head.[58]

Besides roofs, any decent-sized wood was prized by the mines to reinforce their underground shafts and tunnels. Some wood ended up in the boilers of steamers, which gobbled up no less wood in the late '30s than they did when they were first launched almost 60 years earlier. The rest went to firewood, a commodity that, by the early '40s, had become so rare around Yellowknife it was being barged into town during the summer or in winter, sledded in from far away by dog teams or horses. These services were duly reflected in the price of firewood — almost $25 a cord. In 1993 dollars, this would be the equivalent of about $300. The idea of someday switching to oil as the primary source of heat was becoming less of a joke around town.

In less than a decade, newcomers riding shield country's first wave of industrial development brought profound changes to the landscape. At the same time, the landscape left deep-seated marks on many of the newcomers. Most arrived in the pioneer spirit, smelling nothing more than a resource frontier yet to be carved out. This was the "New North," a mineral empire ready to be vanquished. Writing in the summer of 1939, Jock McMeekan, mining editor of Yellowknife's first newspaper, the *Prospector*, portrayed the region as a promised land of hidden wealth, predicting that "this Great Slave Lake Country will be a great, rich productive adjunct to the prairies of the west." According to his vision, Yellowknife would serve as the "spearhead of development" in this "great practically unexplored empire."[59]

Visions were common in those days. Edgar Laytha had quite a colourful one as he toured the Con Mine site still under construction in 1939:

> And in this new world I suddenly thought of — the Middle Ages. I thought of turreted high castles of medieval conquerors in subjugated territory. The great fortress-like mine castle of the Con, which looks like a symbol of modern winged knights upon the conquered Northland, reminded me forcibly of the castles of days gone by.[60]

A new footprint on the northern shield landscape: Consolidated Mining and Smelting mine ("Con Mine" for short) situated on the south shore of Yellowknife Bay. It is parked over a massive gold-bearing shear zone hosted in Archean-aged volcanic rock. *(1949, Busse /NWT Archives)*

When the plane door slammed shut behind them or the boat chugged away from the dock, most new arrivals had less grandiose visions. Rather than conquerors, they more often felt like prisoners. Their sense of isolation was most pronounced during breakup or freeze-up, when trips to and from "the Outside" were impossible.

> During the six to eight weeks' spring break-up they will be shut off from the world. There will be no way of escaping from their bush camp, no way for planes to reach them while the lakes are filled with floating ice blocks. . . . If anybody comes down with appendicitis during break-up, it will be too bad for him. No doctor can get near him.[61]

Seen from the trail, early prospectors and adventurers often described the land as naked, endless and dreary, choked with soggy muskeg and blood-thirsty bugs. But it took on a different face when they stopped to smell the flowers.

> At last the path leads out of the muskeg up steep, rocky hills, which may be pregnant with gold inside, but outwardly are dreary enough. The sun burned them hot, dry, dead.

. . . Yet even in this dreary country you meet pleasant travelling companions. Wild orchids, smaller of blossom and less gay than their tropical relatives, spring up here and there in the wet moss. Orange-yellow flowers greet you, and you are astonished at the green-cupped, dwarf relatives of our lilies. Singly, in clusters and in masses you see wild roses; they have a strong perfume, and blossom luxuriantly, climbing half-way up the hills to form delightful pink islands.[62]

In one of the first issues of the *Yellowknife Blade,* Jock McMeekan wrote of his special affection for the shield landscape northeast of Yellowknife:

It is one of the most beautiful parts of the Yellowknife District. The hills rise to a height of five to seven hundred feet, separated by wide, flat bottomed valleys, covered with a white carpet of caribou moss. There are huge straight spruce trees, wide-spaced and no underbrush. It is all so clean.[63]

Some, like Vic Ingraham, a Yellowknife prospector and entrepreneur, found more than pretty flowers and aesthetic refreshment. In the "epic grandeur" of shield country he found a deep, soul-nurturing contentment:

Here in the North, the good North, man need not armor himself against man. Here the little man becomes the great man. In the North the waves of the lakes are cruel, but the waves of life are merciful. They carry Man to ever higher ideals of pure humanity. They carry him nearer to God.[64]

Laytha described the first round of Yellowknife's boom times as "the most modern and most efficient gold-rush of all time. . . ."[65] The second boom came in 1945, brought on by a combination of good science and good luck. Legendary findings of gold had been delineated at the new Giant Yellowknife mine. A year earlier, Neil Campbell discovered the massive gold-bearing shear zone which has kept the Con Mine in business ever since. Time for more postwar gusto.

Whoosh! The flood gates were pushed open and a tidal wave of people invaded Yellowknife. The summer of 1945 was the wildest, maddest of them all. Peace had come to Europe. Men were coming home. The world relaxed. Yellowknife boomed! Shutters were now off all the buildings.[66]

The aromas that drifted in through those open windows were not all pleasant. The official 1945 population of Yellowknife was 3,000, with most of them crammed into what now is called Old Town. Prospectors who came back to town after several weeks or months in the bush could either bathe in the lake or do without. Many did without. No one had plumbing. Besides, the lakefront was severely polluted.

The town had mushroomed during the past two years, tents giving way to jerry-built shacks, without plan or system. . . . Most of the buildings were of lumber shipped from the south, hastily erected and unpainted. The most intrusive of these were the privies, which in fact fairly dominated the scene. Perched on rocks or spongy turf, they were always in sight in almost every direction. So jumbled were the dwellings that a privy in one man's backyard fairly blocked another man's front doorway.[67]

In the mid-1940s, Yellowknife was no more than a glorified refugee camp for southern transplants. (Is it still?) Human crowding was great for business, especially the hotel business. At peak season, up to 100 guests were put up in the 14-room Yellowknife Hotel, with a quarter of them sleeping on cots in the beer parlour. But it was bad for the environment. And then, as now, this meant that it was ultimately bad for the humans.

Not surprisingly, the Yellowknife Hotel was one of the biggest polluters. Its wastewater spewed out of a four-inch pipe directly into Back Bay, the primary source of most people's drinking water. Add to this broth a cascade of wastes from all the cafes and most of the privies, and it seems a miracle that a serious epidemic never broke out.

They came close. On March 15, 1947, a Con Mine employee died of typhoid fever, a disease directly attributable to foul water. An emergency typhoid hospital was set up in the local Legion Hall. Some wastewater pipes were swung 180° away from Back Bay towards Yellowknife Bay. Privies were shored up with fresh peat. Eventually, new bylaws were passed prohibiting altogether "the escape of liquid wastes." The tank truck industry

No room on the Rock — by the late 1940s physical congestion and related environmental health problems forced Yellowknife to expand "up the hill" (upper right). This 1956 view looks south over Old Town, with Back Bay to the right and Yellowknife Bay to the left. *(Busse/NWT Archives)*

boomed as wastes were hauled away to a safer distance. After all this, two more people contracted typhoid. They survived.

As Yellowknifers were despoiling their nest, the long arm of southern civilization continued to influence the shield hinterland. For instance, some historians believe that the rise in human numbers was directly related to a significant local decline in caribou. "One of the effects of civilization's encroachment into the Yellowknife Bay area was to drive the caribou from the scene — a pity but inevitable."[68]

Gone were the days when "unbroken columns of caribou" blackened the ice along the north shore of Great Slave Lake. Towards the end of the 1940s, it became clear that caribou were once again on the decline. Like population trends before and since, the exact cause may never be known. Shifting migration patterns, calving success, winter range changes due to fire, disease and parasites — these and other factors provide grounds for endless speculation. Increased hunting pressure was the explanation favoured by John Kelsall, one of the foremost caribou biologists of the day:

A 1958 postcard depicting the large concentrations of caribou that once gathered in the vicinity of Yellowknife. *(Busse /NWT Archives)*

The caribou crisis of 1949-55, when reduced to the above essentials, presented no mysteries. It was evident that one major factor, human kill and crippling loss, was causing annual population deficits. Systematic and range-wide manipulation of the human kill was the only tool necessary to increase, decrease, or stabilize the caribou population.[69]

Forest fires, human-caused and otherwise, continued to march across the northern shield. Mining companies became increasingly nervous about the dwindling supply of timbers left to prop up their mines. Airline companies — and passengers alike — learned to live with the reality of prolonged groundings due to smoke. Hunters and trappers became convinced that diminishing game populations were directly linked to recent conflagrations. Water bombers and fire crews became big-ticket items on the federal government's northern budget. A "broad publicity offensive" was launched to help curb the spread of fires. A quip from the federal government's *A.B.C.'s of Forest Fire Protection* captures the spirit of its entreaties to the public:

G is for Game
The Forest's its home.
When the Forest is burned
It has no place to roam.[70]

Against this backdrop of environmental change, yet another development boom hit, this one fuelled by uranium. This was the second time a distant war had touched the face of shield country. During World War II, the region had a brief fling with tungsten, a material used to make armour-piercing shells that, it was hoped, would drive back Rommel's Panzer tank divisions in Europe. Demand for tungsten reached sufficient levels for a short-lived mine to dig up some concentrated ore found on Outpost Island in the middle of Great Slave Lake's East Arm. During the last half of the 1950s, the Cold War was heating up and uranium, used to build American atomic and hydrogen bombs, quickly came into vogue.

In the first three months of the 1955 staking season, the Mining Recorder in Yellowknife filed 4,600 mineral claims — that was up from 400 from the previous year. On June 1, 1957, the region's second uranium mine (after Eldorado on Great Bear) went into production, Rayrock Mine, 90 kilometres north of present-day Rae-Edzo. Although the mine operated for only two years, its radioactive tailings remain an environmental concern for the local Dogrib even to this day.

Besides uranium, much gold and other valuable minerals were discovered during the staking boom of the '50s. Several new mines opened up as a result, including Tundra Mine, northeast of Yellowknife beyond the trees, and Discovery Mine on Giauque Lake. Along with the prospectors and miners getting off the plane at Yellowknife and Fort

Smith was a growing cadre of foresters, soil scientists, tourist outfitters and government bureaucrats interested in assessing the region's broader resource potential.

As the northern resource frontier continued to roll back, the region's transportation network rolled forward. As of 1948, it became possible to drive an all-season road from Edmonton to Hay River on the south shore of Great Slave Lake. By 1964, the finishing touches were being put on a railway to Pine Point. In 1960, the highway, such as it was,

Looking out the mouth of the Indore mine southeast of Great Bear Lake in the spring of 1954. This was one of several mines in the region to go into production as a result of the '50s staking boom. *(Busse /NWT Archives)*

leapt across the mighty Mackenzie River at Fort Providence (via ferry or ice bridge). In 1961 it rolled into Yellowknife. If Prime Minister Diefenbaker had got his way, it would have carried on around the East Arm of Great Slave Lake, past promising mineral reserves near Fort Reliance, and connected back to Hay River. This was just one of Diefenbaker's "Roads to Resources" meant to roll through the great untapped northern shield.

A catwalk spanning the Yellowknife River in 1960. Though a bridge and all-weather road extending eastwards was constructed soon after this photo was taken, the road — now called the Ingraham Trail — stopped far short of Diefenbaker's grand vision of an East Arm loop. *(Busse /NWT Archives)*

For better or worse, depending on which way you look at it, the idea for this road was scrapped. But this was not to stop boom number four. The road brought development costs down to more attractive levels. The region was in a heyday of exploration and expansion on all fronts. The population of Yellowknife grew to 7,000. According to the town's satisfied mayor, Ted Horton, "The sound of hammer and saw was on the land." So was the sound of lawn mowers. Kentucky Blue Grass had found its way onto the northern Canadian Shield, growing in the front yards of ranch-style bungalows found in any other "civilized" North American community. The New North was becoming sophisticated.

Along with that sophistication came increasing public concern for environmental health issues created by a quarter century's worth of gold mining. Since 1938, mines on one side of town or the other had been roasting impurities out of the gold ore and sending them up stacks into the air or down pipes into the water.

Arsenic was the element that caught most of the public attention, perhaps because of the Shakespearean drama that glamourized it as an easily administered form of poison. Over the years, thousands of tons of the stuff had been released into the environment and now it resided, in varying concentrations, in the water, in the lake sediments, even in carefully nurtured cabbages growing in backyard gardens.

By the early 1960s, there were signs posted around Yellowknife and regular notices in the local paper warning residents not to melt snow as a source of drinking water. Rumours among the local Dene band claimed that two children, whose families relied almost exclusively on meltwater during the winter, had died of arsenic poisoning. A battery of studies begun a few years later — and still going, off and on — revealed that, yes, there were some amazing levels of arsenic in and around Yellowknife. That was the bad news. The good news was that these levels were on the slow but sure wane, thanks to improvements in the technology of arsenic recovery. A decade or two from now, there may be healthy (even edible) fish stocks in the town's several lakes, which today, practically speaking, are off limits for fishing and swimming. Drinking may have to wait a little longer.

A curious by-product of gold mines is that emissions from their roasting stacks, particularly sulphur dioxide, changed the landscape in ways that improved the prospects for finding more gold. Some geologists have gone so far as to applaud the denuding effects of these toxic emissions:

> Pleistocene glaciation, recent forest fires and acid rain from the roasting of arsenical gold ores has cleared much of the plant cover, soil and lichens from the outcrop. . . . The excellent exposures provide an opportunity to study volcanic structures and textures in detail.[71]

For at least three decades, impurities in the ore and chemical residues from milling processes were routinely spilled into adjacent waters or spouted into the air from Yellowknife's two gold mines. Though environmental controls have improved significantly since this 1955 shot of Giant Mine was taken, chronically high levels of SO_2 emitted from the roasting stack continue to stress downwind vegetation, particularly lichens. *(Busse /NWT Archives)*

Aside from minerals, the region's fish and wildlife became caught up in the renewed spirit of frontierism that characterized the 1960s. Many politicians in the fledgling territorial government were beginning to see the region as a magnet for tourists and big-game hunters and maintained that wildlife was not being used to its "full economic advantage." One member of the Legislative Assembly stated this sentiment rather strongly: "In the modern Arctic there is no species of animal, not one single species, being exploited to the highest degree possible for the maximum benefit of the Northwest Territories." [72]

Countering the push to exploit northern wildlife was a growing movement in southern Canada to view such places as shield country as among the last great wildlife refuges in North America. Many people, primarily southerners, harboured sentimental notions that wildlife should be left alone completely.

Ever since the '60s, the business of managing the region's wildlife has required a delicate political and biological balance between commercialization and conservation. During the late 1970s and early '80s, management decisions came down heavily on the conservation side, since the consensus of biologists was that barren-ground caribou herds were headed for another tailspin. There was even talk of possible extinction, as there had been 100 years earlier. New techniques of population monitoring soon dispelled these fears by revealing caribou numbers that far exceeded those of previous surveys.

At the tail end of the twentieth century, most of the region's wildlife populations are enjoying a period of relative abundance. The caribou are as unpredictable as ever. During the winter of 1991, they came right through Yellowknife and groups of a dozen or more loafing animals often were spotted on the lake ice within sight of City Hall. Hunting close to the city became so intense that a one-kilometre no-fire zone was imposed along the only road out of town. A year later, the caribou wintered far to the east, forcing hunters to carry an extra can of gas or two on their snowmobiles.

Barren-land populations of muskoxen, which were brought to the brink of extinction by hunters from both sides of the tree line, have made a dramatic recovery since the turn of the century. So abundant are they in the southern portion of their range that the Chipewyan community of Lutsel k'e (Snowdrift) on Great Slave Lake is exploring ways of reinstating its traditional muskoxen hunts on a regular basis. [73]

Forest fires continue to march across the taiga shield landscape, occasionally threatening towns, roads and power lines, as they always have since such things came north. Water — so much water! — continues to pour through countless lakes and rivers that remain nameless still. Long, cold winters and, some say, the bugs of summer continue to impose constraints on the peopling of this landscape, which may well contribute to the long-term preservation of its wildness.

Many who only pass through carry on the long tradition of frontier exploration, now most fervently expressed in the quest for glittering jewels buried in the rock. In the fall of 1991, a startling discovery of gem-grade diamonds was made by Point Lake near the tree-line fringe of shield country. Almost overnight, a region once equated with gold mining became an international Mecca for corporate pilgrims gripped with diamond fever. Not since the 1930s has such a flurry of prospecting activity swept over this landscape. As of July 1993, over 16 million hectares — an area about twice the size of Scotland! — had been staked in the quest for diamonds. Another boom time cometh.

Meanwhile, unaided by diamond drills or dynamite, the Precambrian bedrock underfoot continues to weather and wash away, as it has for billions of years. Grain by grain, the seemingly immutable stage on which shield country's dramas have played themselves out is disintegrating at an average rate of one millimetre every 100,000 years. In a million years, this landscape will have lost one centimetre of its crystalline skin. In a million years, what soils, what plants, what creatures, what people will be dancing on this slightly lowered stage? Whatever or whomever they are, they likely will be no less wondrous than the cast of characters to date.

Today's Landscape

Introduction

While surveying large chunks of Yukon wilderness as prospective recreational parks, I once had the unusual experience of doing a reconnaissance flight in the company of a multidisciplinary team, the likes of which has probably never flown together before or since. On board were a geologist, botanist, wildlife biologist, ornithologist, archaeologist and bat ecologist. With noses glued to the aircraft's windows, each of us looked down upon the same landscape, yet to each it was entirely different. The geologist saw rock formations. The botanist saw vegetation communities. The wildlife ecologist saw animal habitats. The ornithologist saw bird habitats. The archaeologist saw likely sites for prehistoric camps. I never did get what the bat ecologist was looking for.

Whether scientist or soothsayer, we all see landscapes through different filters and mental projections. Having looked long and hard at the taiga shield, I have come to understand it in relatively simple terms: first by its main components, such as climate, soils, plants, birds and animals and then by its habitats, common associations of the components that can be easily recognized in the field. Part 1 of this section, "Threads in the Environmental Fabric," introduces the primary ecological threads colouring this landscape. Part 2, "A Tapestry of Habitats," shows how these threads are woven together.

Part 1:
Threads in the
Environmental Fabric

Here in the northern forest we can see the direct effects of physical factors on organisms, we can unravel the simplified food web and examine the component food chains, we can see and experience directly the effects of seasonal changes in light. A number of ecological principles are put on display in graphic clarity. In the taiga, students of ecology can easily grasp the fundamental concepts of the science as they are laid bare around them.

— William Pruitt, *Wild Harmony* [74]

8

Climate: Wind, Water and Ice

Something snapped last night. In the middle of the fifteenth night of August, the seasons pivoted. Summer was sent reeling by the year's first thunderstorm. In the snug shelter of my unlit home from a north-facing window overlooking Yellowknife Bay, I watched the drama unfold — a good night not to be camping.

It was the lightning that woke me up. It set the night sky on fire with dragon-shaped forks and rippling sheets of electrostatic light. With each flare, I could make out the squall line bearing down out of the northeast. At its leading edge was a hulking, cylindrical cloud that rolled forward, like a tornado lying on its side. Behind it was an armada of ominous blue-black clouds, seemingly held aloft by thunderbolts.

Along the lakeshore, tall willows swayed and shuddered violently, leaning away from the wind as if to escape. On a whim, I phoned the Yellowknife weather office to hear their taped message for 3 a.m. An anonymous voice told me that a gale warning was in effect, with winds gusting as high as 80 kilometres per hour. After a few hours of this kind of wind, waves out in the middle of Great Slave Lake can rise up to five metres, the height of a modest two-storey house — a good night not to be boating.

Watching a thunderstorm at night through binoculars is invigorating, though the strobe-light effect can be a little hard on the eyeballs. At the peak of the storm, aim anywhere and something interesting is bound to come into view: a writhing bolt of lightning, a wall of whitecaps on the lake, the illuminated inner anatomy of a cumulonimbus cloud or, as on this night, a pair of red-necked grebes.

There they were. With thunder exploding directly overhead, buckets of rain flying sideways through the air and lightning flashing all around, these two grebes bobbed idly in the shallows. I lowered my glasses, not really believing what I was seeing. More lightning. The grebes were swimming nonchalantly *away* from shore, out into the froth. I lifted my glasses, aiming into the darkness at the spot where I figured they would show up next. Another tremendous flash, this one simultaneous with a deafening thunderclap — may my roof be spared! No grebes. After five more minutes of this peculiar hopscotch approach to bird watching, I concluded that either those grebes got fried by a 200,000 ampere, 10,000°C pillar of fire or they simply took advantage of the extra light to do a little late-night fishing for sticklebacks or fingerling pike.

By morning all was calm. Dripping willows, flattened sedges and a lane full of puddles were the only evidence of the storm. Yet something told me that an irrevocable change had occurred overnight. Was it the uncanny stillness, the cool bite in the air, or the stark Himalayan blues of a rain-washed sky? It's hard to put a finger on it, but I wasn't the only one in town to notice. "How *was* your summer?" people started asking after that storm. Somehow we knew. There's a magic time in mid-August when the robust green fullness of summer gives way to fall, a time of ripening, letting go and decay.

Over the next four weeks, the reddening cranberries, now hiding a lime-green crescent below and a tart, snowy centre within, will flush crimson throughout and fill the air with a musky sweetness. Cattail reeds, now tipped with a thin arrowhead of brown, will fade and wither like dying flames of green. Poplars and birches, now tinged with occasional yellowing leaves, will form brilliant rivers of peach and gold through the boreal forest as photosynthesis shuts down for the winter. Diving ducks such as mergansers and scoters, now gathered in restless rafts on the big lakes, will make their exodus for southern shores. And beneath the ducks, the layered thermal structure of lakes, now warm up top and cool below, will break down completely as surface waters cool and sink. All these events and countless others are triggered by this part of the planet tipping away from the warming, life-giving rays of the sun.

By the fall equinox, September 21, night regains its grasp on northern skies. Flickering stars, rippling northern lights, rich evening sunsets — all those nocturnal events we forget about in the summer are back on the celestial stage in full force. For a brief period in the planet's annual cycle around the sun, night and day are held in perfect balance, with each allotted exactly twelve hours. Above or below the Arctic Circle, on the equator or the Tropic of Capricorn, it's the same 50-50 split. Light-wise, shield country is put on an equal footing with the rest of the world. So why at the fall equinox is it 40°C plus in the tropics while temperatures up here may be below zero? And for that matter, in the summer when we're getting almost twice as much sunlight as at the equator, why do our thermometers rarely rise above 20°C?

High noon during the subarctic winter. Low-angled sunlight provides little energy to the earth's surface. *(John Poirier)*

It's another case of quality over quantity. Throughout the year, equatorial regions are bathed in a daily infusion of intense solar energy. As the sun swings directly overhead, tracing a loop oriented 90° to the plane of the horizon, its rays traverse a relatively short column of air, resulting in minimal diffusion and filtering of heat and light. In boreal regions, solar radiation is less direct, arriving obliquely at a much lower angle of incidence — if it arrives at all. The net effect is a spreading of the same radiation over a larger surface area; it is less concentrated, less intense. A flashlight beam tipped from 90° above the floor to 45° or less provides a simple demonstration of this profound effect.

To make matters worse, at least for fair-weather types, a low angle of incidence means that radiation striking the upper Northern Hemisphere must travel through relatively more air before it reaches the ground. The sun's energy is therefore further reduced by greater reflection and refraction from water vapour and airborne dust, much like the scattering of light in a murky fish tank.

The combined effect of these two phenomena is a distinctly northern energy handicap. It's no wonder then that, while Venezuelan coffee beans and pineapples ripen in the early October sunshine, the million lakes of shield country are freezing over.

Frost crystals grow out over black ice. (*John Poirier*).

By the middle of October, day length has dropped to less than ten hours. At this time of year, over six minutes of sunlight are shaved off each day around Great Slave Lake. Above the Arctic Circle, this rate of loss is significantly greater, hastening the land's descent into a sunless winter.

On still mornings, fog hangs over open stretches of rivers and the quickly closing lakes. Ptarmigan and barren-ground caribou cross over the tree line seeking shelter and forage in the boreal forest. The dying light, subzero temperatures and freezing lakes drive out the last ducks, geese and swans. The poplars and birch stand naked against chilling easterly winds. Bedrock cracks and fissures are dusted with snow.

In summer, the winds of shield country prevail from the south, flung out from the northern arms of the continental air mass. This is a zone of generally low atmospheric pressure that spans the continent's interior and spins counterclockwise like a giant pinwheel. Twirling clockwise over the barren lands and arctic islands is the arctic air mass, a large climatic region dominated by high-pressure weather systems, which are typically cold, dry and stable. Like the caribou, this air mass migrates south of the tree line in the fall. As it moves, it displaces the warmer, wetter continental air mass to the south. The belt of convergence, or "front," between these two air masses is highly unstable, creating a line of battle between high- and low-pressure systems. As these systems sally back and forth across the skies over shield country, the weather becomes ridden with storms bringing the easterly gales of September and October and the heavy snows of November and December.

By January, the arctic air mass has won the battle — as it always does — having bullied the continental air mass well south of 60°. Snowstorms are less frequent and the air can be bitterly cold. The ice on Great Slave Lake is now a metre thick, enough to support a D7 bulldozer weighing close to 20,000 kilograms. The sun shines little more than five hours in Yellowknife and much less farther north. Above the Arctic Circle, the sun has slipped completely out of sight. On those sunless "days" of winter, dawn merges unbroken into dusk in a brief waxing and waning of muted light.

On still, cloudless nights, reflected light from the moon or stars alone illuminates the landscape — enough light for sled dogs and skiers to navigate down a winding trail, enough light for owls and lynx to hunt for snowshoe hare. But on such nights, there's usually little stirring in the boreal woods. To limit radiant heat loss to the clear night sky, warm-blooded creatures must take cover beneath an insulating shield: a thick canopy of spruce boughs, a blanket of snow, or a cabin roof. On such nights, the land gives off the stored energy it receives by day from incoming infrared rays. With no clouds to reflect some of this outgoing energy back to earth, it flows unimpeded into the infinite heat sink of outer space.

The three ingredients for a first-rate sundog display are low-angled sunlight, a dusting of cirrostratus clouds and cold temperatures in the upper atmosphere. This sundog was captured by Henry Busse over Yellowknife's Old Town in 1952. *(Busse/NWT Archives)*

When night temperatures dip below −40°C and winds are nil, people in northern communities often wake up to discover that the air has turned to ice — crystals of ice formed when warm, moist exhaust from vehicles, homes and office buildings meets the dry, frigid air. The resulting ice fog hangs like a shrouded dome over town, trapped in a temperature inversion by the denser cooler air above. Warmer temperatures and wind above 10 kilometres an hour break down the inversion layer and cleanse the air of built-up moisture and pollutants.

Under conditions of extreme cold, ice crystals can also form in the upper atmosphere, creating a halo of scattered light rays around the sun — if it's shining. Why it's called a sundog I'm not sure, but the effect can be enchanting. Veil-like cirrostratus clouds, 10 kilometres up, bring out the best in sun dogs, since they often consist of the kind of ice crystals that cause the greatest refraction or bending of light rays: six-sided tubes of ice floating straight up and down in the air. If a sundog is especially bright, it will glow red on the side closest to the sun, blue on the far edge and yellow in the middle. At sunrise or sunset, this effect may be enhanced when the halo is doubled like a rainbow or sliced through by a brilliant shaft of refracted light.

Some winters the arctic air mass refuses to budge. Clear skies and −40°C temperatures can last for weeks. In town, conversation about the weather peaks during such cold snaps, the main point of discussion being: "When will it end?" Those putting more faith in science than speculation often turn to Environment Canada's weather office hotline for a hopeful glimpse of tomorrow's weather (they claim to be correct 71 percent of the time[75]). On one frigid day in late January 1990, the phone in the Yellowknife weather office rang almost off its hook, with a record 1,680 calls coming in from a curious public. During another more dramatic cold snap 80 years earlier, the entire Western Arctic was gripped in −50°C temperatures for several days over the Christmas season. By New Year's Eve 1910, most thermometers had bottomed out. As the clock struck midnight, Fort Good Hope logged in at −61.7°C, the current record low for the Northwest Territories.

The weather breaks when chinks in the well-armoured arctic air mass are hit broadside by incursions of warmer, low-pressure air from the west, south or east. The battle between pressure systems resumes and the winds begin to howl. Temperatures may rise suddenly, sometimes above zero. There's nothing like the respite of a mid-winter thaw to raise human spirits. Icicles forming and actually dripping in early February is a marvellous thing to behold after two weeks of −40°C weather. The woods temporarily come alive as ptarmigan, snowshoe hare and red squirrels emerge from their snowy shelters. Like an abruptly ended dream, it's usually all over in a day or two. Plunging temperatures and vicious wind-whipped snowstorms soon send everyone ducking for cover once again.

On a still day −20°C is downright balmy. Combine this temperature with a 20 kilometre an hour breeze — about enough wind to extend a good-sized flag — and conditions become perilous. The wind chill effect creates the heat loss equivalent of 39°C, cold enough to freeze exposed flesh in three to five minutes. Still holding at −20°C, double the windspeed to 40 kilometres an hour and you hit the equivalent of −51°C. Climatologists have an official term for this condition: "bitterly cold." That's one notch away from "extremely cold," when "outdoor conditions are dangerous!"[76]

At windspeeds over 30 kilometres an hour, powdery snow tends to become airborne, creating a possible whiteout, during which the land, sky and horizon all blend into an obscure uniform whiteness. Needless to say, navigating through such treacherous conditions is next to impossible.

Besides creating hazards to life and limb, winter winds reshape the architecture of snow, both at the surface and at a microscopic level. In exposed areas such as lakes and rock outcrops, strong winds sculpt loose snow into striking patterns of drifts, ripples and swells. The jumbling and tossing of wind-borne snow crystals wears down their delicate rays. Reduced to shattered fragments, these crystals nest more closely together, resulting in a hard, dense layer. If this layer is only on the surface, it is known as wind slab. Deeper reworking of snow by the wind can transform it into a semi-solid mass strong enough to easily support a fox, wolf, caribou or human.

Deep in the boreal woods, where the effects of wind are negligible, snow undergoes a different kind of transformation. Throughout most of the winter, the soil below the snow is usually much warmer and moister than the air above it. Heat and moisture flow upward from the soil to the upper layers of cooler, drier snow. Through the process of sublimation, molecules of water break off from the deeper snow crystals and are carried upwards until they condense and refreeze themselves onto colder crystals above. Over the

Dense, hard-packed snow sculpted by unobstructed winds blowing over Great Slave Lake. *(John Poirier)*

Though seemingly still and changeless, snow crystals undergo an unceasing process of metamorphosis once they fall to the ground. For instance, sublimated by heat given off from the earth, the basal layer of snow may disappear completely over the course of the winter, particularly in sheltered situations such as this dense spruce forest. *(John Poirier)*

winter, the lower crystals of snow get smaller and weaker, while the upper ones get fatter and stronger. Eventually the basal layer of snow is reduced to tiny, round ice crystals with the consistency and firmness of sugar.

This so-called corn snow is pretty mushy stuff to walk on. Sometimes the basal layer is completely eroded away through this process, leaving, just above the soil, a vacant space with fragile, latticelike walls and roof. While snowshoeing or skiing in late winter, I have often been spooked by the sudden and far-reaching collapse of these under-snow caverns.[77]

By the vernal equinox on March 21, the days at the latitude of Great Slave Lake are once again twelve hours long. Daylight now *increases* at a rate of about six minutes a day. The big thaw is on. At first, it is limited to the warmest part of the day, when only the upper snow surface melts. At night, this soggy snow freezes hard, becoming a tough, ankle-biting sun crust.

As the days get warmer, the temperature gradient in the snow cover flips upside down. In the winter, it was warm below and cold up top; in the spring, it is warm up top and

Candle ice, a springtime treat for the eyes and ears. *(Bill Braden)*

cold below. This results in a further transformation of snow structure as water drains down through symmetrical channels and the upper snow crystals turn into long, vertical cylinders called ablation needles. The effect is best appreciated as lakes and rivers begin to break up into candle ice, accompanied by the musical tinkling of these prism-like needles.

It takes a long time for winter to finally release its grip on the land — six weeks or more from the time melting begins to the final disappearance of snow. And it's perhaps another month after that before the big lakes give up the last of their ice. The main reason is that in the spring as much as 85 percent of the solar energy striking snow and ice cover is reflected directly back into outer space. Most of the energy that does get through is used up in the processes of melting and evaporation. It seems almost a shame, all that radiant energy pouring into shield country, yet little of it available for heating the air or the land.

This climatic arrangement does, however, have an up side. For a few weeks in May, the land lies freshly exposed and free of snow, while the big lakes remain covered in ice thick enough to walk on. The hiking this time of year is fabulous. Walk just about anywhere, enjoy shirt-sleeve temperatures, witness the returning birds — and no bugs! An annual "birdwalk" is held this time of year, during which a parade of bird-watchers from Yellowknife marches across the blue-green ice of Prosperous Lake to a favourite wetland alive with ducks. On the way, I often pause to marvel at the shimmering heat waves and bizarre mirages hovering above the ice: islands float over the horizon, clouds quiver below it, spruce trees stretch like rubber bands into the sky. Such phantom images give visual expression to the extreme reflectance properties of ice.

More magic happens in the first weeks of June. After eight months of dormancy, the swollen buds of poplar and birch trees finally burst open, unfurling delicate flames of green. The forests come alive with the singing of warblers, kinglets and thrushes. Around lakes and wetlands, bald eagles cruise the shoreline in search of fish for their recently hatched offspring, dabbling ducks let loose the final passions of courtship, and loons practice their yodelling without restraint. The prevailing winds begin to blow again from the south, as the continental air mass regains its supremacy over much of shield country. Each day, the sun marches farther north across the sky, until it reaches the summer climax of its journey on the June 21st solstice, the celebrated longest day of the year.

For the months of June, July and August, Yellowknife holds the distinction of being Canada's sunniest city, boasting an average summer total of 1,065 hours of bright sunshine. On an average June day, the birds start singing around 2:30 a.m., when the sun comes up. They shut down around 10:30 p.m., when the sun makes a hesitant dip below the horizon.

Although light is abundant this time of year, the warmth provided by infrared radiation is still in short supply. This is due to the consistently low angle of the sun's rays *throughout* the year, not just in winter. As a result, Yellowknife's average summer temperature is only about 14°C, with the record high being a modest 33°C recorded on July 16, 1989. By comparison, Regina, Saskatchewan, receives almost 150 hours *less* sunlight than Yellowknife over a summer, yet its average temperature for this season is around 18°C and its record high is a searing 43.3°C. The all-time high for the entire Northwest Territories goes to Fort Smith at 39.4°C on July 18, 1941.

North of the Arctic Circle, a summer "day" may last from 48 hours to several weeks, depending on latitude. In spite of the apparent glut of sunshine, summer temperatures are significantly lower than in Yellowknife. Why? Partly due to the yet shallower angles of incidence. Even at high noon on June 21, the sun swings no more than 45° above the horizon at the Arctic Circle. This angle continues to decrease dramatically with increasing latitude. Another reason is the greater influence of the arctic air mass, which in the summer is parked just north of the tree line. Finally, a tremendous amount of the energy that does get through is expended on burning off last winter's snow and ice.

It is ironic that the sunniest season in shield country is also the rainiest. Half of the region's precipitation falls during the months of June to September — one-third of the

Multiple exposure of the midnight sun crossing the horizon, Port Radium, June 1946. *(Busse/NWT Archives)*

year. August is the wettest month; rain one day in three is the norm. On such days, while waiting out a twelve-hour downpour in the confines of my tent, it becomes hard for me to believe that this region is the cold "desert" it is claimed to be. Yet by climatological reckoning, it is comparable to the northern reaches of Tibet, a cold, windswept place that receives only 150 millimetres of precipitation a year. Many places on the "dry" prairies, such as Edmonton and Winnipeg, receive over twice as much precipitation. Prince Rupert, on the British Columbia coast, gets fifteen times more.

They can have it. Despite the taiga shield's scanty precipitation, moisture is plentiful — in lakes and rivers, in the muskeg, in the frozen ground. Cold temperatures limit evaporation and the air's ability to hold moisture. Vast areas of relatively flat topography limit runoff. As a result, the precipitation that does fall, either as rain or snow, is stored for a long time before it eventually cycles back into the clouds.

All that stored water, in turn, can modify climate on a local level. In particular, large bodies of water such as Lake Athabasca, Great Slave and Great Bear can act as giant heat reservoirs, wind generators, precipitation sources and cloud enhancers. Water heats up more slowly than land and holds its heat longer. These lakes therefore have a general moderating influence on adjacent lands, slightly delaying the arrival of green-up in the spring and postponing the arrival of frost in the fall. For example, Yellowknife, perched on the north shore of Great Slave Lake, averages 111 frost-free days a year, giving it a longer growing season than landlocked Regina, almost 2,000 kilometres to the south.[78]

During stable periods in mid-summer, winds on the big lakes follow a predictable daily rhythm well known to sailors and fishermen: "In by day, out by night." Onshore sea breezes are set in motion when strong daytime heating causes warm air to rise over the land, sucking in cooler air from over the water. At night, gusty offshore land breezes often develop in the opposite direction as the land cools and warm air begins to rise over the water. As this air rises, it usually takes a moisture load with it, contributing to local cloud cover and, eventually, showers.[79]

August is anything but stable. The eternal battle between major pressure systems begins again in earnest. With the inevitable storms comes the rain. During this month alone, the region on average receives 43 millimetres of rain, representing almost one-fifth of the year's total precipitation. In August 1969, a record rainfall of 142 millimetres fell on Yellowknife — over half the yearly quota.

During a blockbuster thunderstorm in August, temperatures may drop suddenly as leaden clouds gush forth their huge payloads of moisture. The wind unbridles its pent-up fury, while lightning plays against the backdrop of a darkening sky. For me, such storms represent more than a turning point between seasons. Over the space of a few hours, the storm's many faces embody the violent pulsations in light, energy and precipitation that characterize shield country's climate through the cycle of a year.

9

Soils and Permafrost:
Drunken Forests and Boiling Mud

Those old farming stories from north of 60° seem to belie the apparent limitations imposed by climate — cabbages the size of basketballs weighing five and a half kilograms; fields of sweet clover averaging over a metre in height; a single garden plot yielding enough potatoes to feed several communities for the entire winter.[80] Though true, few of those stories come from the taiga shield. Most come from the Western Arctic's big river valleys — the Mackenzie, the Liard, the Slave and the Hay — which form corridors of relatively warm summer air and whose banks are covered with thick till and rich nutrients left by several millennia of spring floods. The climate of shield country is too harsh and its soils too poor for our cabbages to make it anywhere near the Farming Hall of Fame.

For many indigenous northern plants (cabbages are indigenous to the balmy Mediterranean coast), the shield edge represents a distinct geographic barrier beyond which life becomes intolerable. East of this edge, search forest and bog for snowberry, silverberry or honeysuckle bushes, for the water hemlock, virginia cherry or wild ginseng. Chances are you won't find them. Even though some of these species extend north beyond the Arctic Circle or grow high in the Mackenzie Mountains, they have shunned the taiga shield. Species that do survive here have had to adapt to one of the most important factors limiting life on the shield: poorly developed soils that are thin, cold, coarse, sour and soggy.

Whether rich or poor, soil is a miraculous substance. It is itself alive and breathing. Within the subterranean world of sand, clay, silt and organic matter are hundreds of different kinds of soil organisms co-existing in a complex web of life. There are tiny carnivores with such curious names as harvestmen, pseudoscorpions and sun spiders. There are phytophages, the plant eaters, such as nematodes, insect larvae and various beetles. Then there are the main players, the saprophages, eaters of dead and decaying matter, such as millipedes, mites, bacteria and fungi.

Together, these organisms perform the many chemical and biological transformations that provide the foundation of life above the soil: the physical breakdown of plant and animal matter, the conversion of fine organic material into soil nutrients, such as phosphorous, potassium and nitrogen, and the alteration of those nutrients into forms that can be absorbed and ultimately recycled by plants. All of the main groups of soil organisms and the vital functions they perform are represented in shield soils. It's just that, compared to more benign environments, their numbers are low, their functions slow and the raw material they have to work with pretty skimpy.

While living in south India in the early 1980s, I had the satisfaction of planting a few rows of corn, watching its prodigious growth, enjoying its sweet harvest and witnessing its tough, fibrous stalks turn to dust — all in the space of six months. There a stout rosewood branch falls to the forest floor and within a year or two is converted into a barked tube of crumbly red soil. Here, well-preserved spruce logs criss-cross the taiga shield like spilled matchsticks, having fallen to fire sixty years ago or more. Near the tree line, I once discovered some weathered old spruce stumps showing rough-hewn gouges suggesting the work of stone tools. Were they chopped down during a time, way back, when metal implements were scarce? Or are they simply the more recent marks of hard-luck but resourceful campers who forgot their axe? Either way, these stumps spoke to me of the plodding rate of soil development north of 60°.

Whether in tropical India or subarctic Canada, the particular kinds of soils that exist anywhere are an expression of complex interactions among climate, moisture conditions, vegetation and parent material — the mother rock from which soils are born. Like all living things, soils follow a developmental cycle of their own: they are created, they mature, they become more complex, they disappear.

The agents that help create soil — water, wind, fire and ice — are ultimately the same ones that destroy it. Flowing water creates soil through the gradual deposition of suspended sediments removed from landscapes upstream. Wind acts in the same way, creating here, destroying there. Forest fires hasten the creation of soils through the quick release of nutrients bound up in vegetation. They can also destroy soils, burning them down to cinders, which are in turn carried off by wind and water. The most powerful agent of soil change and development in the Northern Hemisphere has been glacial ice. Like cosmic bulldozers, continental ice sheets up to five kilometres thick and a continent wide dumped soil in abundance on some landscapes. In others, next to nothing was left behind. In this respect, the taiga shield was a definite loser.

Over much of this area, the Pleistocene ice age pushed the process of soil development back to square one. Most of today's thin soils began literally from scratch about 8,000 years ago, when the last wave of glaciers retreated to the northeast, leaving in its wake a sea of scoured and pitted bedrock or a boulder-strewn blanket of gravelly till. Soil development since then has been exceedingly slow due to the cold temperatures and short summers, which severely restrict the activities of soil-building organisms and reduce the productivity of plants, the main fodder for those organisms.

The uniformity of this region's climate, topography and parent material is reflected in the dominance of two main soil types: cryosols and brunisols.[81] The word "cryosol" derives from the Greek *kruos*, meaning frost, which reflects the distinguishing feature of this cold-climate soil: ice. Cryosols are permeated with ice in different forms, from tiny crystals or thin, barely perceptible ice layers to huge ice lenses or veins several metres thick. The particular ice conditions in cryosols vary widely, depending on the texture of their individual soil particles — for instance, fine silt versus coarse gravel. On the taiga shield, cryosols are found most often in peaty muskeg areas, where little heat penetrates underground even during the prolonged days of summer.

In whatever form or location, cryosols share another common feature that sets them apart from other soils. They are regularly churned by cryoturbation, a process that disrupts the soil through countless cycles of freezing and thawing. As a result, most cryosols show no hint of the standard soil layers, or "horizons," common in most other types of soil.

Most northern brunisols, the region's second main soil type, do show at least a loosely defined pattern of horizons, though they too are regularly churned by frost action. Up top is the dark brown "O" horizon, or organic layer, which contains mostly decaying plant matter and the millions of soil organisms that break it down. Below this is the "A" horizon, a dark mixture of decomposed organic matter and mineral silts. Because of acidic conditions, the upper soil layers are subject to intense leaching, a process that moves valuable soil nutrients beyond the reach of plants through the downward percolation of water.

Ironically it is plants — conifer needles and some mosses — that make a big contribution to the acidity of northern soils. Organic compounds of potassium, nitrogen and phosphorous, being chemically basic, become soluble in acidic water and are washed out of the soil. Many important inorganic compounds such as iron and aluminum also go into solution, sometimes accumulating in the lower, finer-grained "B" horizon. Such soils may show a brownish-red tinge created by the high concentration of iron oxides leached from above. Put simply, the soil is rusting. Beneath the soil is the parent material, usually represented by granitic rock, or some close geological relative, which also sours the soil because of its inherent acidity.

Besides being thin, cold, acidic and nutrient poor, northern soils throw another challenge at plants trying to make a living on their surface: they are often underlain by permafrost. Defining this term turns out not to be as simple as the name at first implies. Back in the 1960s, R. J. Brown, one of Canada's leading experts in the field, stated once and for all that "permafrost is not permanent."[82] However, since then, whole dictionaries on the subject have been published, thick with nomenclature describing perma-this and perma-that. The other twist is that neither "frost" per se nor ice is necessarily present in permafrost; it may be bone dry, containing no frozen moisture. The British have another term that provides a closer match with reality; they call it "perennially frozen ground." According to conventional scientific wisdom in Canada, what's down there is this: soil or rock that remains at or below 0°C for at least two years.

So defined, permafrost underlies no less than one-fifth of the planet's land surface — its high mountains, its polar regions, and much of the subarctic boreal forest. In Canada, that figure jumps to one-half. On the taiga shield, the distribution of permafrost is patchy, or discontinuous. Around Lake Athabasca, less than 30 percent of the land is underlain by permafrost that averages less than 10 metres in thickness. Around Great Slave Lake and north to the tree line, that percentage ranges anywhere from 30 to 80 percent, with thicknesses approaching 100 metres. Beyond the tree line, permafrost becomes continuous, underlying 100 percent of the land and reaching depths of up to a kilometre in the High Arctic.

Permafrost occurs whenever the annual amount of heat going into the ground is less than the amount that goes out, a net loss situation that leaves the ground at below zero temperatures the whole year through. In southern parts of the taiga shield, where permafrost is sporadic, such temperatures occur only in places where the soil is buffered from the warm summer sun: in low, boggy areas having a thick insulating layer of moss and peat, in dense spruce stands providing heavy shade to the forest floor and on north-facing slopes and stream banks where the sun shines only a few hours a day.

Northward, permafrost takes hold in an ever wider array of vegetation types and exposures until the entire landscape, above and below, experiences average temperatures that are below zero. This corresponds to a mean annual air temperature of about −8.5°C. (For comparison, Yellowknife's is −5.4°C.) At this point, permafrost becomes continuous — it underlies everything. With steady subzero temperatures in the soil, permafrost

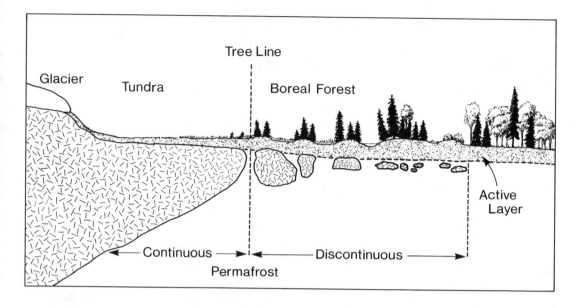

Permafrost varies in thickness and distribution depending on vegetation cover and latitude. At the taiga's southern fringe, it occurs sporadically, reaching depths of only a few centimetres below the active layer. In the high arctic, permafrost is continuous and may be half a kilometre thick.[83]

grows downwards much like tree rings, as a thin layer of frozen ground is added each winter. This process continues, year after year, until an equilibrium is reached where cooling from above is balanced by heat flowing up from the earth's interior. The thick permafrost found in the continuous zone represents thousands of years of incremental growth that might have begun as far back as the last ice age.

To dig up some shield soils, hoping to get a good look at permafrost, could be frustrating. If you're lucky, you may discover thick veins, wedges or lenses of ice mixed with the subsoil that usually indicate the *presence* of permafrost. More often than not, there's nothing much to see. At −5°C or +5°C, most soils look pretty much the same. Try digging through permafrost with a shovel and the difference becomes clear, at least to your muscles. Where permafrost makes its presence more plainly known is in its expression on the surface.

Whether sporadic or continuous, thick or thin, the business end of a permafrost terrain is its spongy top layer, the mantle of soil that thaws every summer and freezes solid in the winter. This is the active layer, which varies in depth from a few centimetres in the High Arctic to over a metre along the southern fringe of the discontinuous zone. This layer is often quite soggy in the summer because the precipitation that enters the soil — what little of it there is — has a tough time leaving. It can't percolate down and drain away completely because permafrost is impermeable to water. It can't turn around and evaporate very quickly because this process is severely constrained by cold temperatures. It is the damming effect of permafrost and the regular freezing and thawing of the active layer that create the unusual surface patterns and shapes so distinctive of permafrost terrains.

Such features are best enjoyed from an airplane cruising about 1,000 metres above the barren lands. From there you might see intricate webs of irregular polygons shaped like giant cells on a skin of tundra, a latticework of shallow, rectangular ponds suggesting well-manicured rice paddies, tonguelike lobes creeping down a hillside, or linear stripes and craterlike circles etched into a lifeless plain of shattered rock. If you were to fly south to the tree line, you would see these patterns gradually subside, giving way to a terrain characterized by raised plateaus of scattered spruce surrounded by expansive sedge meadows and humpy bogs. Flying above the thicker forests farther south, you might notice that the larger trees seem somehow askew, leaning at random angles, creating the impression of a "drunken forest."

Back on the ground, you can see other permafrost features that are common throughout the taiga shield. For instance, when walking over the peaty, organic soils that overlie most muskeg, who hasn't set foot on a hummock? These are bulging, oval-shaped mounds usually one to two metres in diameter. Hummocks usually have a silt or clay core whose fine, closely packed soil particles offer little space for ice crystals to move into as they expand. When the active layer freezes each winter, there is no place for the soil to go but up, like frozen milk forcing up the cap on a milk bottle. Aptly called frost heaving, this is the same process that creates havoc with building foundations and roadbeds throughout the North.

Hummocks come in all shapes and sizes depending mainly on the soil's moisture content, its texture and the thickness of the active layer. In some hummocks, the disturbance caused by frost heaving is so intense that the roots of any plants attempting to grow on them are torn apart. These are mud boils usually found in coarse-textured glacial till. Also called circles, they are round patches of bare soil that appear to have bubbled up from below.

A hummock of around 30 centimetres or less is called a tussock or, as my young daughter would say, a baby hummock. Created by plants such as sedges and cotton grass, these often carpet entire wetlands in high densities and, when combined with deep puddles and pools, make normal strolling a challenge. Much larger mounds in the muskeg are palsas, solitary hulks of peat and layered ice that can rise above the general level of a bog to a height of 7 metres and measure 100 metres across.

These lumps and bumps on the northern landscape did not go unnoticed by early tourists in shield country. Casper Whitney, an American adventurer and journalist, left this account of a snowshoe trek through muskeg in the early winter of 1896:

> Level footing is exceedingly scarce, the wind blows the snow 'whither it listeth', and you cannot know whether you are about to step on top of one of those innumerable mounds or into one of the many gutters that cross-section the swamp. You know after you have taken the step. Nine times out of ten you land on the slanting side of the mound, and slip and trip and turn your ankle and use yourself up generally.[84]

Permafrost poses different kinds of problems for plants. Besides physical disruptions caused by frost heaving and soil movement, permafrost is hard on plants because of its subzero temperatures and impermeability to water. Its chilling effect delays warming of the soil in the spring and keeps it relatively cool all through the summer. This affects plants profoundly by slowing their growth rate and general metabolism. Its damming effect results in waterlogging and poor aeration of the soil above, leaving plant roots short on essential oxygen. Certain chemical reactions and activities of soil-building organisms are also restricted by low oxygen levels, causing further impoverishment of soil nutrients.

The net effect of permafrost on vegetation is clearly negative — all this on top of the inherent limitations of taiga shield soils! On the other hand, the effect of vegetation on permafrost is largely positive. Its main effect is to shield permafrost from degradation by solar heat. A thick mat of moss and peat offers the best thermal protection. Disturbance of such cover by fire, encroachment of water or a few passes of a bulldozer can have profound, sometimes disastrous, consequences on permafrost soils. For example, on the arctic slope of Alaska, trails made by winter tractor trains in the late 1940s caused slumping and erosion of the underlying permafrost soils, resulting in gullies that today are 10 metres wide and up to 5 metres deep — and they might be growing still.

Besides such pronounced local effects, the thermal sensitivity of permafrost is such that small fluctuations in a region's climate produce subtle changes in the extent, thickness and temperature of permafrost. In the discontinuous zone, this can result in the gradual disappearance of permafrost on some sites. Evidence of "fossil" polygon features well south of Lake Athabasca point to such changes in the distant past. Since its initial formation, permafrost in any area may have vanished and reformed several times during periods of climatic warming and cooling.

Held in delicate thermal balance by a complex array of ecological factors, permafrost is one of the most pervasive and dynamic expressions of the ever-changing landscape of shield country.

10

Plants: At Home in the Kingdom of Spruce

Plants are fascinating. Whether you know their names or not, give them your focused attention and you will discover a wonderful diversity of colours, shapes, textures, smells and, for many species, taste. The process of getting to know the unique character and life story of a particular species can transform a green blur on the forest floor or water's edge into a familiar acquaintance with whom a lifelong relationship is established.

Who among plant lovers does not have a mental list of favourite species that capture the senses and stir the emotions in some special way? The balsam poplar is high on my list of shield country favourites. Its generic name, *Populus*, is derived from the Latin word for "people." This is a tree of the people, a distinction harking back to Roman days when poplars ornamented public squares reserved for the masses. Here they ornament the wild shores of subarctic rivers and lakes. Its specific name, *balsamifera*, refers to the tree's sweet-smelling resin given off by its leaves. Its buds, too, are fragrant. Squeeze one after the leaves drop in autumn, at −25°C during a snow storm, or in springtime as it swells with imminent unfolding: the perfume is there. Mixed in breathable pouches with rose petals and fragrant herbs, its buds are sold commercially to add an exotic scent to drawers full of lingerie. When it comes to abundance of annual growth, this tree is unmatched. In a region where plants seem bound by one law of growth — low and slow — the balsam poplar's yearly eruption of slender twigs and fat, glistening leaves is a source of reassurance in the potency of spring.

Then there's the twinflower, another favourite of mine and of Karl Linnaeus, the 18th-century botanist who took it upon himself to classify every living thing. The father of modern-day taxonomy, Linnaeus described and named plants from around the world. He published over 180 books on the subject, the thickest of which was entitled, matter of factly, *Plant Species.* Of the tens of thousands of species he examined, only one did he hold so close to his heart as to give to it his own name: *Linnaea borealis.* Found on the floor of spruce forests in his native Sweden and throughout the circumpolar world, this diminutive shrub trails through the moss, bearing two — always two — sweet-scented, nodding flowers, which range in colour from creamy white to a rosy purple. For me, as perhaps for Linnaeus, it is the twinflower's unassuming elegance and pleasing symmetry that make it so endearing.

Each plant species has its own charm, each has its own unique role in holding together the fabric of life on the taiga shield. Collectively, the plants of this region share one thing in common: they have what it takes to survive the trials of a subarctic climate, with its violent pulsations in solar energy, its burden of long-lived snow, its desert-dry air, its abrasive and heat-robbing winds and its impoverished soils underlain by permafrost.

The flora of shield country is relatively simple, dominated by a few types of plants, many of which are aggressive, wide-ranging and ecologically undemanding. The black spruce, for instance, is not picky about where it takes root; it is found on exposed rocky slopes, in well-drained valleys or in wet lowland bogs. Other species succeed by adopting strategies that allow them to populate the rawest, harshest habitats where all but the hardiest species are excluded. The most accomplished of these plant pioneers are lichens.

Lichens specialize in living where nothing else can, on naked, sun-scorched rocks, tree bark or lifeless twigs and on thin, barren soil or sterile sand. And together with a handful of moss species and rugged dwarf shrubs, lichens form the bulk of what is commonly called tundra, that ubiquitous blanket of vegetation covering Canada's most hostile environments. Lichens need very little to live on: light, moisture and a place to anchor — that's it. Unlike most other plant forms, lichens can suck moisture and nutrients directly out of the air. Consequently, with no need for roots, they are free to grow on hard, impenetrable surfaces such as Canadian Shield rock. Studies of large lichen colonies on the shield suggest that some lichens, through unbroken generations, have held onto a specific patch of rock for thousands of years. These colonies represent a possible direct living link to the first colonizers that moved in at the end of the last ice age.

One reason that lichens have such a competitive edge over other plants in hostile environments is that, strictly speaking, lichens are not plants at all, but fungi and algae growing together in a mutually beneficial or symbiotic partnership. When, through evolution, these two unrelated organisms merged into one, a new life form was born that

Foliose lichen at home on the rocks. (*John Poirier*)

was clearly greater than the sum of its parts — the nearest thing to a closed and self-sustaining system that the living world has produced.

The fungus component produces the lichen's fruiting structure, its skeleton, which provides an anchor and catches and stores water — both of vital importance to the survival of algae on land. Living independently, fungus cannot make its own food. In the form of mushrooms and moulds, fungi requires dead or decaying material to survive. But in lichen, fungus food is supplied by photosynthesizing algae, which, safely encased in the fungal body, create starches and other complex compounds out of the very air we breathe.

Most lichens can reproduce vegetatively, that is, without the dance of sex. A trotting moose, a driving rain or the errant footfall of a curious naturalist can easily send many pieces of lichen flying. If a single fragment of the fruiting body, called the thallus, falls onto favourable habitat, it can start an entirely new lichen colony.

Such propagation by disintegration is most likely to occur while lichens are in their crunchy, brittle state. When light or moisture becomes scarce, lichens shut down most of their life functions, they shrivel up, their colours fade. They are able to bide their time in true dormancy until conditions improve. The switch back to full vitality is most dramatic after a long-overdue rain. Lichens then become swollen and supple as seaweed. Their colours deepen slowly, yet almost perceptibly. Black lichens turn a dark olive, white ones turn a powdery green, orange ones glow like hot coals. Tiny photosynthetic engines return to full power, driving the lichen's spreading growth front along at a lively pace of about two to four millimetres a year.

As a group, lichens show astounding tolerance to the harsh demands of life on the shield. This is due as much to their shared survival strategies as to the habitat preferences of individual species. Ironically, many species have very restrictive survival needs, making them intolerant to subtle changes in rock chemistry, moisture, microclimate and microtopography.

Some species thrive on volcanic rock and shun anything to do with granite; for others it's vice versa. Some species prefer sunny south-facing rock surfaces; others can't handle the greater light intensities and do best facing north. Some species have a relatively thick skin and are not bothered by the wind-driven snow that typically bombards exposed sites; others can survive only in more sheltered cracks and crevices, where chances of wind abrasion are nil. This diversity of habitat preferences creates the striking tapestries of multicoloured and multitextured lichens that are so distinctive a part of the shield "look."

Between the 60th parallel and Great Bear Lake, the taiga shield is home to around 350 species of lichens, a daunting number for amateur bryologists to master.[85] A good way to come to terms with such overwhelming diversity is to first recognize the three main growth forms in which lichens are packaged.

Crustose lichens, as the name implies, form a thin, uniform crust or film whose entire under-surface clings tightly to the substrate below. Try in vain to peel off a piece of this kind for a closer look and you'll understand the intimacy between lichen and rock. Unlike the crustose lichens, foliose lichens are more loosely attached to their place of growth and are usually more complex in design. Their surface is organized into elaborate lobes or scaly fronds, giving them a leaf-like appearance (*folium* being the Latin word for "leaf"). They often grow in a rosette pattern, forming targetlike mats on tree trunks and rocks. Most prevalent in this part of the world are the fruticose lichens that have a shrubby or hairlike appearance. They are made up of tiny branches that may be relatively simple, looking like elfin fingers or trumpets. Or they may be extensively divided, giving the impression of a beached colony of delicate corals or sponges.

Where lichens give way to forests, taiga vegetation can be characterized by one word — spruce. Constituting 90 percent of what European explorers called the "forest of little sticks," spruce here comes in two models, one black, *Picea mariana*, the other white, *Picea glauca*. As distinct as their names might imply, these species are often difficult to tell apart, especially when they grow side by side, each assuming the same shape, that of a gigantic pipe cleaner. In the words of boreal ecologist James Larsen, both black and white spruce "may occupy a bewildering array of habitats and otherwise behave in an unaccountable manner."[86]

The standard account portrays black spruce with drooping, rather straggly branches and a compact, narrow crown. The best place to find them is said to be in moist, acidic habitats such as bogs and cool, north-facing slopes. White spruce, on the other hand, are often described as fuller, more stately trees that prefer dry slopes, riverbanks and other well-drained sites.

While all this is true, you can expect to find much overlap in the shape and habitat preferences of these two species. The definitive difference between them is the presence of tiny rust-coloured hairs along the twigs and new growth of black spruce and their absence on white spruce. This clue becomes obvious with a magnifying glass or a good set of 20-20 eyes at about two centimetres. Having neither, I use another foolproof technique, yet to be patented, which relies on a quick look at their cones. On the black spruce they are about the size and shape of mothballs and their cone scales are thin and slightly fringed at the end. White spruce cones are longer and more robust, with scales that have a tough, even edge. Knowing the difference between these trees is important to me since they are, after all, the dominant life forms around us.

Why is it that spruce and other needle-bearing trees are so at home in cold climates everywhere? Some botanists say this represents the outcome of a battle that's been going on for over 100 million years between conifers and the flowering deciduous trees. Near the end of the Mesozoic era, deciduous trees went through an explosion of diversification and spread rapidly over the face of the planet. Until then, conifers had dominated the world's forests, since way back in the Palaeozoic era, but now they faced stiff competition from the deciduous upstarts. Today, deciduous trees are many times more abundant on earth than conifers. The theory suggests that conifers have been driven back to harsher strongholds, like the Subarctic, where hardwoods cannot easily follow.

Whether it is because conifers are in exile from warmer lands or that they are simply cold-hardy opportunists, the fact remains that they dominate the world's largest forest, the boreal forest, and it is their needles, devised over 300 million years ago, that allow them to retain this dominance.

By hanging onto their needles all year, conifers get a headstart on spring compared to other species that spend the winter naked. When things warm up, rather than expending precious energy and time on the production of leaves, conifers can begin photosynthesis and growth right away, thanks to their ever-present needles loaded with chlorophyll. Even in winter, if a warm spell arrives and sunlight is sufficiently intense,

Spruce — the dominant life form of the taiga shield. Chances are it is white spruce on the higher, well-drained ridge and black spruce on the lower, soggier ground. Only close inspection of their cones or young branches can verify their true identity. *(John Poirier)*

photosynthesis can be switched on temporarily. This process is given a boost by the heat-trapping property of needle-covered branches that capture incoming solar radiation, bounce it back and forth and eventually absorb a good proportion of it with only a small part lost back to space. Also, the needles and dense twigs catch the wind and slow it down, so that the effect of windchill is much reduced in a conifer forest compared to a more exposed stand of deciduous trees. In these ways, conifers miss no opportunity to exploit the scant energy trickling down to them.

As far as needles go, the tamarack, *Larix laricina,* is a bit of an oddball. In summer, this feathery tree could easily go unnoticed, blending in against a sombre green backdrop of spruce. Come the fall, these trees stand out like candles as their spirally arranged needle clusters turn a brilliant gold — and then fall off. In the spring, a completely new set of needles is produced but not until long after the snow melts and the air warms up. As a result, at about the time tamaracks are just resuming photosynthesis, spruce have already grown several centimetres and their reproductive activities are in full swing. It's therefore no wonder that this species is a minor player in forests of the taiga shield. Low in numbers and patchy in distribution, tamaracks are relegated to marginal habitats in the wettest of lowlands. Without an all-season cloak of energy-trapping needles, tamaracks are poor competitors in the cold domain of spruce.

For most conifer species, needles bestow other advantages that are unrelated to climate. For instance, many insects that infest the tender leaves of deciduous trees find needles too tough to pay them much attention. That's not to say that conifers are invulnerable to insects; many conifer forests south of 60° have been utterly ravaged by the dreaded spruce budworm, a pest just beginning to make its mark in this region. In the face of this formidable enemy, the more distasteful a tree's needles, the better. Here the lowly black spruce is least in danger. Because of the extremely high acid content of its needles, this species comes last on the budworm's list of preferred food items, trailing well behind balsam fir and white spruce.

Having highly acidic needles, the black spruce has a way of fouling its habitat for all but itself and a few acid-tolerant shrubs and mosses. Over time, the gradual upward growth of moss can pose a threat to the slow-growing black spruce, engaging it in a constant struggle for rooting space.

Again, the black spruce has a handy survival tactic. It's called layering. As the tree's spreading lower branches become overtaken by moss, they respond by putting down roots. From each of these points of contact, a new stem grows upward. The result is a crowd of spiky shoots coming up through the moss around a central trunk, like a candelabrum. One such spruce found in the muskeg of northern Quebec had 37 stems, the tallest of which was only seven metres high but almost 100 years old.

The black spruce, white spruce and tamarack are all members of the Pinaceae family — the pines. But the only true pine on the taiga shield is *Pinus banksiana*, the jack pine. Like most plants, it has another Latin species name (depending on your favourite botanical authority), *divaricata*, which refers to one of the tree's main diagnostic features: its pair of twisted needles that diverge or spread away from each other. It has another common name too: the scrub pine. Why scrubby? When the jack pine grows on thin, poor soils, like most up here, its trunk is often contorted and its branches gnarled and knobby — "a sorry-looking tree," according to one popular tree guide.[87]

To me, jack pines seem more stalwart than sorry as they stand up resolutely against all the shield's undiluted rawness. Looking like antique bonsai trees, carved and pruned by stiff winter winds, they clutch tenaciously to minute clefts on an unyielding bedrock skin. They find soil in the driest, most exposed sites where other trees cannot root or in thin, porous sands where desiccation is a constant threat. Another accomplished pioneer, this species often grows in pure stands, springing up in the wake of forest fires. Many of the cones on a typical jack pine will open only when exposed to the kinds of temperatures caused by a fire — around 50°C or more. Thanks to this peculiar reproductive strategy, the next generation of jack pines often begins life in a nutrient-rich layer of ash. In appreciation of the jack pine's distinctive look or its rugged lifestyle, the public recently voted it in as the Northwest Territories' official tree.

As well adapted as the conifers are to the rigours of northern life, they must share some of their turf with a handful of non-coniferous species that have impressive adaptations of their own. For instance, the trembling aspen, *Populus tremuloides*, is one of the most widely distributed trees in North America, stretching from Mexico to the Beaufort Sea. It has no problem with intense dryness, wind and cold and will live in a wide variety of soils. This is a fairly exhaustive recipe for success. But in the boreal forest, perhaps its most useful adaptation is its ability to send out suckers.

The trembling aspen has the option of reproducing without investing energy in the production of seeds or facing the many perils of germination, such as late spring frosts or hungry rodents. They do this by sending out a fast-growing mat of spreading roots from which abundant young stems or "suckers" are thrust up into the sunlight. Given optimal habitat — sandy, well-drained ridges that slope to the south — a few large trees can produce enough suckers to populate an area of land the size of a football field. The resulting aspen stand is not really a group of individual trees but a colony of clones, replicas fostered from the same genetic stock.

The ability to reproduce vegetatively permits aspens to rapidly colonize areas denuded by fire, adding vigorous patches of deciduous trees to the boreal mosaic. If another fire wipes out the colony soon after it is established, new sprouts will arise from the roots of killed trees. Exposed to regular fires, an aspen stand can regenerate itself this way for centuries. But if a colony reaches full maturity before another fire comes along, it will in one generation succumb to its greatest foe: its own shade.

The aspen's suckers are intolerant to shade and will die off after a few years if their light supply is cut by the trembling canopy above. In effect, the aspen stand eventually becomes a death bed for its own kind and a nursery bed for seedlings of other species that thrive in shade, particularly spruce. Investigate any mature aspen forest in shield country and you will likely see young spruce trees dotting the forest floor, vanguards of doom for their smooth-trunked stewards. On such sites, the forest may change more or less completely back to spruce in a matter of decades. While in their prime, trembling aspens add richly to the boreal storehouse of food resources for wildlife. Their twigs are favourite winter fare for moose, their buds for grouse and ptarmigan, and their bark for snowshoe hare and beaver.

The light-loving, fast-growing trembling aspen softens an otherwise rugged landscape dominated by spruce and rock. *(John Poirier)*

The birches share a love of light and are often found in the company of aspens. The so-called paper birches are actually a complex group of closely related variations on a theme. This situation creates something of a nightmare for plant taxonomists. I own three comprehensive plant references, each of which I have sworn exclusive allegiance to at one time or another. One reference splits this group into three distinct species, calling them the paper birch, the Alaska birch and the western birch. Another recognizes two species, one of which has two subspecies. My current botanical bible, Porsild and Cody's *Vascular Plants of Continental Northwest Territories, Canada*, includes only one species, calling it simply the paper birch, *Betula papyrifera*, and recognizes two formal varieties.

Seeking reason within this apparent confusion, some botanists say that over thousands of years the northern birches have adapted to the waxing and waning of glaciers and other climatic fluctuations by developing an ability to interbreed freely with one another, even with quite distant relatives. By maintaining a diverse and free-flowing gene pool, the birches are able to exploit an assortment of habitats while keeping their options open for future changes in climate. The effects of this genetic flexibility are most obvious around the tree line, where rampant interbreeding between birch trees of the taiga and birch shrubs of the tundra has created hybrid swarms of plants that display every combination of size, shape and texture imaginable.

Something about the classic paper birch (subspecies and varieties aside) has given it a central place in many legends, poems, paintings and songs. Its power of inspiration may come from the pristine whiteness of its bark, the graceful curves of its slender, purplish twigs, or perhaps its many gifts to those who make their home in the boreal forest.

A Dogrib trapper from Rae expressed to me his appreciation for birch in clear and simple terms: "The wood is strong and fast" — strong for making snowshoes, axe handles and paddles; fast for making sled runners and toboggans. Its wood splits easily and gives off considerable heat even when green. Like the sugar maple, its spring sap can be boiled down into a delicious syrup. But the greatest gift of the paper birch is its bark.

For early peoples in the region, birch bark provided one of the main building blocks of their culture. It's amazing stuff — light, flexible, strong and waterproof. It peels off easily. And since birches grow not only in pure stands but also mixed in among the spruces, it is abundant. I once joined a Dene elder named Elizabeth Mackenzie in the woods on a search for birch bark. I watched her pass up dozens of trees as she used her well-trained eye to look for a clean, unblemished specimen. Once she found it, all it took was a few quick slashes and pries with her knife and the bark seemed to pop off the tree. Mackenzie's ancestors used birch bark to make all the utensils needed for food preparation. They ate off birch-bark plates. They stored food and other supplies in birch-bark baskets that were, and still are, decorated with floral designs made from porcupine quills. They rolled it into tubes and called moose to their death. Their children played with birch-bark toys. And they travelled the shield's many waterways in birch-bark canoes.

"Canoe birch" is what they call it in some parts of eastern Canada where the biggest trees grow and the biggest canoes were made. During the height of the fur trade, canoes up to ten metres long with a load capacity of four tonnes were coming out of the forests of New Brunswick and Nova Scotia, each made from the bark of one or two trees. In the

Shield Country

GALLERY

Bald Eagle launching from spruce-top perch. (*Paul Nicklen*)

Yellowknife — an urban island amidst a sea of subarctic wilderness. (*Jamie Bastedo*)

One of shield country's many unnamed lakes;
this one shaped like a stocky woodsman with an axe poised over his shoulder. (*Rene Fumoleau*)

Looking like petrified dinosaur dung, these volcanic pillows near Yellowknife's Giant gold mine formed over 2.6 billion years ago deep beneath a Precambrian sea. (*Jamie Bastedo*)

Precambrian inscriptions etched in volcanic greenstones. The raised portions have eroded more slowly than the surrounding rock due to localized silicification. (*Jamie Bastedo*)

A sheetlike granodiorite dyke fills an ancient fracture in the darker volcanics of Ranney Hill north of Yellowknife. (*Jamie Bastedo*)

Proterozoic-aged cliffs along the north shore of Redcliff Island on Great Slave Lake's East Arm. (*Jamie Bastedo*)

Stromatolite fossils of blue-green algae. Blanchet Island, East Arm of Great Slave Lake. (*Jamie Bastedo*)

Gold is mined from a massive shear zone along the West Bay Fault, clearly visible through the middle of this scene. (*Jamie Bastedo*)

A classic roche moutonée showing glacial abrasion and striations on the upstream side (left) and an abrupt, plucked surface on the downstream side (right). Whaleback in rear. (*Jamie Bastedo*)

A pronounced glacial groove ploughed into the bedrock by a boulder embedded at the bottom of the glacier. (*Jamie Bastedo*)

Fine beach sand sorted by the waves of Glacial Lake McConnell. Bristol Pit, Yellowknife. (*Jamie Bastedo*)

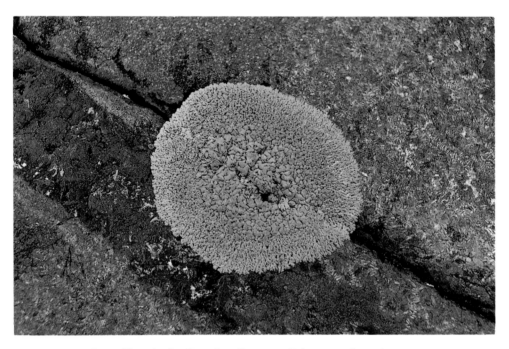

One of hundreds of species of crustose lichens — plant pioneers of the bald Precambrian Shield. (*Jamie Bastedo*)

A community of fruticose lichens — an elfin world of delightful colours, textures, and shapes. (*Jamie Bastedo*)

A solitary paper birch reaches its autumn finest by mid-September.
The trembling aspens behind it may all be genetic clones, having sprung
from the same mat of spreading roots. (*Jamie Bastedo*)

With the arctic air mass locked firmly overhead, the dawn sunshine brings little
warmth to this sheltered bay along the east shore of Great Bear Lake. (*John Poirier*)

Two antlerless bull caribou engage in a brief sparring match. (*Doug Heard*)

A red fox sits alert in a bank of fresh snow. (*John Poirier*)

Ermine or short-tailed weasel. This expert "mouser" lets out a
shrill shriek when agitated or seizing its prey. (*Paul Nicklen*)

The lynx's enormous feet help it float over deep snow while pursuing its main prey, the snowshoe hare. (*Paul Nicklen*)

A spruce grouse treads lightly over spring snow. (*Paul Nicklen*)

A 130-year-old jack pine, veteran of a forest fire that levelled the surrounding forest in the early 1970s. (*Chris O'Brien*)

Fire the destroyer — a crown fire consumes an old-growth spruce forest. (*Bob Gray*)

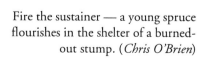

Fire the sustainer — a young spruce flourishes in the shelter of a burned-out stump. (*Chris O'Brien*)

A large beaver dam at the end of a narrow bay creates a rich wetland ecosystem behind it. (*Jamie Bastedo*)

Yellow mastodon flowers, mare's tails, and cattails thrive in a marsh fringing Yellowknife's Niven Lake. (*Jamie Bastedo*)

Sphagnum moss envelopes a caribou jaw. (*Jamie Bastedo*)

The horned grebe, one of the most common water birds breeding on shield country's countless wetlands. (*Laurence Turney*)

Good old *Vaccinium vitis-idaea* or mountain cranberry
— the queen of northern berries. (*Chris O'Brien*)

Rubus acaulis, the short-stemmed raspberry. (*Jamie Bastedo*)

In sheltered, moist conditions, this species of blueberry,
Vaccinium uliginosum, can reach grand proportions. (*Chris O'Brien*)

The cloudberry, *Rubus chamaemorus*, ripens to a
musky sweetness by early August. (*Chris O'Brien*)

Magnificent Parry Falls — a bedrock bottleneck
in the "disorganized" flow of water down the Lockhart River. (*Jamie Bastedo*)

Showy flowers along the shores of rivers and lakes: the water parsnip, *Sium suave* (above) and the river beauty, *Epilobium latifolium* (below). (*Jamie Bastedo*)

Snakelike esker provides shelter for isolated pockets of tree line spruce. Humpy Lake. (*Jamie Bastedo*)

Weather-beaten spruce (right) share the shoulder of this esker with a carpet of ground-hugging tundra shrubs. Caribou trail in the foreground. (*Jamie Bastedo*)

Taiga and tundra plants meet at the tree line. Crimson leaves of alpine bearberry, *Arctostaphylos alpina*, needle-like leaves of crowberry, *Empetrum nigrum*, and puckered leaves of mountain cranberry, *Vaccinium vitis-idaea*. (*Jamie Bastedo*)

Bedazzled birders witness the explosive tide of spring migrants.
Unparalleled refreshment for city-weary tourists. (*Jamie Bastedo*)

With intelligent land-use management, the tranquil,
pristine beauty of shield country will live on. (*Jamie Bastedo*)

The taiga shield — one of the last crown jewels among the
planet's dwindling natural treasures. (*Jamie Bastedo*)

northern taiga, where birches are considerably smaller and more crooked, the biggest canoes were half that size. Not that bigger was necessarily better. The stealthy "hunter canoe" measured less than three metres in length and was light enough to be portaged by a child.

To make the outer shell of a canoe, pieces of bark were sewn together with peeled spruce roots, which have the strength and flexibility of rubber. Finished off with a coating of spruce gum around its seams, the birch-bark canoe was amazingly waterproof and durable, sometimes lasting ten years or more. If destroyed while far from camp, a new canoe could be built in a week or two from paper birches always close at hand.

First cousins to the birches are the alders, another group of transitional species that shuns the shade. These are common shrubs that usually seek light along lakeshores and stream banks, hence the generic name *Alnus*, which in Celtic means "neighbour of streams." Alders can also hold their own in moist young forests, either coniferous or deciduous, where light is not in short supply. In favourable conditions, alders can grow so thick that the best way to negotiate passage is to lean well into them half falling, take small shuffling steps and grit your teeth. Such "alder hells" are, however, a definite boon to any habitat from the perspective of soils.

Attached to the roots of alders are many small nodules of bacteria that specialize in pulling nitrogen out of the air and into the plant. When the leaves, roots and nodules decompose, nitrogen floods the soil, resulting in more vigorous plant growth, including perhaps bigger, thicker alders. Besides adding nitrogen to the soil, alder roots play an important role in binding soils together in unstable situations such as high-energy shoreline habitats, where the erosive action of water and ice can be intense. Alder leaves, the biggest grown in the taiga forest, are grown in abundance each year, making a hefty contribution to the soil's upper humus layer, where all plants first take root.

When chewed to break out the sap, alder leaves can be rubbed on mosquito bites to relieve itching. Early aboriginal cultures made a kind of elixir with alder bark that, when taken internally, treated rheumatism. When applied externally, it helped kill pain and stop bleeding in wounds. The active ingredient behind the alder's medicinal value is salicin, a chemical with the same curative properties as household aspirin (acetylsalicylic acid).

In the whole of the Northwest Territories there are only two alder species — the grey or hoary alder, *Alnus incana*, which has deeply toothed leaf margins, and the green alder, *Alnus crispa*, whose leaves are much less jagged. Of willow species there are almost fifty, most of which grow on the taiga shield. They are a confusing lot.

In the lingo of hard-core botany, there are at least sixty ways to say that a plant is not smooth, that it has fuzz, hair or roughness of some sort. From arcuate to hispidulous, pilosulous to woolly, willows in one species or another show it all. The shape and texture

The elegant paper birch, a tree that takes a central place in the ecology and cultures of this region. (*John Poirier*)

of willow leaves vary subtly from species to species and sometimes, on an individual plant, from branch to branch. Depending on habitat conditions, the same species may take the form of a low, dense shrub or a tall, spindly tree. Willow reproductive organs, the catkins, are often diagnostic, but they are usually short lived and aid identification for only a few weeks in the spring. Although I have befriended a handful of species that are readily identifiable, I call most willows "willows." The finer details of willow science I leave to the professionals.

During my impressionable days as a young graduate student roaming the Yukon wilds, I had the privilege of spending a day on a mountain with Dr. Bill Cody, a man who, ever since that day, has been my definition of a keen, I mean *really* keen, botanist. Here's a man who took along a World War I bayonet on his honeymoon — not to ensure his personal safety but to dig up plants. We talked a little while out in the field. We mostly walked. With his eyes constantly sweeping the ground, he looked for new or unusual specimens, stopping occasionally to dig something up or confirm a plant's identity. I noticed that when we passed by groups of willows, many of them quite alien to me, he hardly batted an eye. At one point, where curiosity got the better of me, I trailed behind to check them out and plucked a catkin-laden branch off of what looked like a particularly puzzling species. Resuming my stride beside the master, I asked him what it was. Without missing a step, he glanced at the spent catkin, gave it a quick twirl between his fingers, then made his call: "*Salix bebbiana.*" "How do you know?" I asked, with all the respect I could muster. "You just know," he replied, grinning.

Moose also have a keen eye for willows. They depend on them, especially in the fall and winter when they browse heavily on willow shrublands — so-called "moose pastures." Willows are also a major food source for snowshoe hare, beaver, muskrat, ptarmigan and grouse. During the Pleistocene, willows fed arctic horses and woolly mammoths. Now, as then, willows are of fundamental importance to boreal ecosystems. After spruce, they constitute the main bulk of their woody vegetation. As a group they have adapted to a wide range of habitats, from the wettest bogs to the driest pine forests. Their annual output of leaves and twigs can be astonishing. A colony of willows cut down one year along trails or roadsides may spring back to head height the next, much to the chagrin of the clearers. If a leafless willow twig is stuck in some moist ground in the spring, it may be a substantial bush by the fall. It is the willow's adaptability and tremendous productivity that make it such a key player on the northern stage.

Long before nylon entered the traditional taiga economy, fishnets were made from willow, from long, pliant strips of its bark twisted together with spruce roots. Rope, dog collars, cooking utensils and knitting needles were made from willow. It was used to buttress canoes, rim birch-bark baskets, snare small game and form hoops for drying pelts. Willow sap was scraped from the tender inner bark and eaten, providing a delicious source of carbohydrates. And, like the alders, willows were used to make various medicinal potions to relieve pain, promote healing and reduce fever. The relatively high salicin content in willows is reflected in their generic name, *Salix*. Many of these uses continue today, including the braiding of catkins — so carefully studied by aspiring botanists — into jewellery for children.

Imagine this country without willows. It would be a much impoverished place, both ecologically and culturally. Now take away also the berry-producing plants. It would be a wasteland. The importance of berries in the taiga shield landscape was not lost to Ernest Thompson Seton, a Canadian naturalist who explored the Great Slave–Great Bear region in 1907. About the ubiquitous kinnikinick plant, *Arctostaphylous uva-ursi*, he wrote:

> It furnishes a staple of food to all wild things, birds and beasts . . . it is one of the most abundant of the forest products, and not one hundred yards from the fort are solid patches as big as farms . . . .[88]

Pygmy and arctic shrews, the smallest of mammals, depend on berries, especially in winter when they draw them from stored troves. So do the biggest of mammals. In the fall, when black bears are fattening up for the cold months ahead, 75 percent of their diet may be berries. As the snow melts in the spring, some returning ducks and geese scoop up overwintered berries when there is little else to eat. For other birds like the northern junco, bohemian waxwing and many sparrows, berries are a major food source for much of the summer and fall. Berries find a place in the diet of traditional carnivores such as marten, mink and weasels. Even wolves and foxes will turn to berries as a nutritional stopgap when hunting success is down. In shield country, most food chains are propped up somewhere along the line by berries.

Why are there so many berry-producing plants in the North? A popular theory among some botanists is that by enticing animals and birds to eat them and passing through their digestive tract, the plant's seeds stand a good chance of sprouting in nutrient-rich droppings. This strategy of seed dispersal gives berry producers a competitive edge over other plants that must take root in the thin, impoverished soils dominating this land.

The kinnikinick described by Seton is one of about a dozen common species found on the taiga shield that are members of the heath or Ericaceae family (pronounced "air-i-kay-see-ee"). Heath species tend to be low and compact, a morphological adaptation to the boreal winter. They spend most of it buried in snow, shielded from the chilling, desiccating effects of the wind. Some species, like the alpine blueberry, *Vaccinium uliginosum*, and the decumbent form of Labrador tea, *Ledum decumbens*, form dwarf shrubs no more than ankle high. Others, like kinnikinick, its close relatives the red and black bearberries, *Arctostaphylous* spp., and the famous mountain cranberry, *Vaccinium vitis-idaea*, are lower yet, prostrating themselves along the ground by way of a slender horizontal stem. Each spring, these stems send up a fresh batch of twigs bearing delicate bell-shaped flowers ranging in colour from yellowish green to a warm, deep pink.

The leaves of many heath species have built-in wind shields to protect their undersides during snow-free seasons. This part of any plant is most sensitive to drying because of numerous gas-exchanging stomata, which open wide during photosynthesis. One of the distinctive features of the heath family is the presence of "revolute" leaves, which are rolled downward along their margins, thus sheltering the lower side from wind. Several species show this feature, including the small bog cranberry, *Oxycoccus microcarpus*, the bog laurel, *Kalmia polifolia* and Labrador tea. The two latter plants have an additional adaptation for blocking wind under the leaf: a dense mat of short hairs, which could pass for a good imitation of fur. To further reduce the loss of moisture, most heath leaves have a waxy or scaly covering and are thick and leathery — hence the name for one species, "leather-leaf", *Chamaedaphne calyculata*. Like the conifers, most heaths retain their leaves over the winter to make the most of a meagre supply of solar energy.

The rose family is the next most important contributor to the taiga's annual berry crop. The wild rose, *Rosa acicularis*, itself does not produce berries; it produces "hips," swollen red fruits, which can be remarkably sweet and delicious but when eaten whole can cause a unique form of discomfort. Beneath the hip's pulpy outer layer lurks a bundle of sharp, hairy seeds, which can create havoc with the lower digestive tract, especially on their way out — hence their nickname "itchy-bums." It's better to first dissect out their seeds or mash them and strain them. Or you could go look for a safer member of the rose family, such as the cloudberry, *Rubus chamaemorus* or the short-stemmed raspberry, *Rubus acaulis*, both of which are found in peaty wetlands and along waterways and can, if conditions are right, produce fruit in abundance.

Most prolific of the rose family's berry bushes are the classic wild raspberry, *Rubus strigosus*, and the Saskatoon berry (or serviceberry, *Amelanchier alnifolia*). These species are accomplished colonizers and can spread quickly into areas levelled by fire or some other disturbance. Once established in moist, well-drained soils, they can grow in thick profusion, successfully crowding out other shrubs or trees that might steal their sunlight. In such patches, the berry picking — by both man and beast (as Seton might have said) — is unsurpassed.

Then there is the Ribes clan (pronounced "rye-bees"), four of which occur through-out most of shield country: the red, black and skunk currants and the gooseberry. Identifying these species takes some practice. They are best studied close up with a berry bucket in hand. Luckily for berry pickers, most currants lack thorny stems. Berries of the most common species are smooth, except for the skunk currant, whose berries have bristles. In contrast, the gooseberry has stems with thorns — quite formidable ones — and its berries, striped like miniature beach balls, carry a small tail at the end, a vestige of the former flower.

Among local berry producers, there are several species that, like many of us southern transplants, have relatively few kith and kin up north. Though loners taxonomically, they deserve honourable mention, having spread themselves widely and producing a sizable proportion of the taiga's annual berry crop. The crowberry, *Empetrum nigrum*, for instance, is a one-of-a-kind plant, being the only representative of its family — the crowberry family, in fact. When bearing fruit, it looks like a cross between a black spruce and a black bearberry. Its extravagant output made quite an impression on the British explorer Samuel Hearne, who, in 1795, wrote of such plentiful berries that ". . . it is impossible to walk in many places without treading on thousands and millions of them."[89] And its range is perhaps outstripped only by the dandelion. The crowberry is found from northern Ellesmere Island to southern Vancouver Island, from Iceland to the Aleutians and westward, around the top of the globe, from Siberia to the heathlands of Scotland.

The high-bush cranberry, *Viburnum edule*, and the soapberry, *Shepherdia canadensis*, are two other taxonomic loners that make a big contribution to the northern berry pool. Superimpose their range maps with the boundary of the North American boreal forest, and you end up basically with one line. Like the twinflower and black spruce, these species are diagnostic of an entire biome that spans the continent. In the taiga shield, they are especially common in moist clearings and open deciduous forests. Here they grow as dense knee-high bushes, which in a good year will bear more berries than leaves.

As an early Oblate priest in the region once said, some berries, like the soapberry and high-bush cranberry, "cannot be relished by a cultivated palate without the addition of sugar."[90] Others, like the Saskatoon berry and raspberry, are, to my mind, best enjoyed fresh off the bush. Collectively, they can be used in countless creative ways. A handful of soapberries and their leaves whipped into a rosy froth makes a great shaving cream if you're hard up for it in the bush. Crowberries fermented with sugar produce a sparkling white wine that rivals the best Californian. Currants sprinkled into batter can liven up the dullest recipes for pancakes, muffins or bannock. The list goes on — sauces, jellies, pie fillings; berry blintzes, berry smoothies, berry sour cream tortes. Whether occupying my plate or still tempting me on the bush, berries are, for me, akin to the raven's call and northern lights and fox tracks in the snow — those things we sometimes take for granted but which draw us ever closer to the heart of shield country and make us want to stay.

11

Mammals: Creatures of Winter

When winter came, there was a great fall of snow and the human heart was troubled. There was so much snow that only the tops of the pine trees were showing and it was impossible to do anything. All the animals wanted to leave for warmer places.
— *Book of the Dene*[91]

For mammals of the taiga shield, winter means deep, sometimes impenetrable snow. It means extreme cold. Sometimes it means severe shortages of food. Rather than merely coping from one year to the next with this seemingly harsh and inhospitable season, the mammals of this region are superbly adapted to its mortal hazards. Leaving for warmer places is not among their marvellous repertoire of adaptive strategies.

Winter is largely responsible for the distribution patterns of many northern mammals. It has determined the nature of many ecological relationships between species. More than this, winter has been a powerful influence in the evolution of species themselves. It has helped shape their behaviour, their diet, their mode of locomotion and their selection of habitat. In many cases, it has shaped the appearance and functioning of their anatomy, such as the size of their bodies, the shape of their extremities, the rate of their metabolism and the nature of their fur. In no other animal is winter's imprint more pronounced than the caribou.

Xalibu is what the Micmac of eastern Canada called them. Early French explorers picked up on this word and it has stuck ever since. It means "pawer" or "shoveller," referring to one of the caribou's most important winter survival skills — digging down through the snow for food with its remarkable feet.

Caribou hooves are actually shaped like a shovel, concave and broad, unlike those found on any other species of deer. With a few swift strokes of its forefeet, a caribou can break through the crustiest of snow and dig a deep feeding crater to access buried lichens and shrubs. The animal's distinctive hoofprint is created by two wide toes shaped like crescent moons. Their edges are sharp and slightly protruding, allowing the caribou to grip ice and hard snow with ease. In winter, the toes are separated by thick tufts of hair that serve as an insulating cushion against the cold. Above the caribou's hoof is a set of metatarsal bones that bends upwards with each step, permitting the "dew claws" to share some of the weight and provide extra flotation over the snow.

As distinctive as the caribou's footprint is the profile of its snout. It is thicker and blunter than most deer. Its shape is due to extra spaces in the nose, which act as heat exchangers, warming inhaled air before it gets a chance to chill the lungs and cooling exhaled air, which may rob heat from the body. Short ears, small tails and compact bodies further conserve heat loss by reducing the amount of overall surface area from which precious heat can radiate. The slender legs of a caribou are equipped with a heat-exchange system of veins and arteries that reduces the amount of heat lost from blood travelling to the extremities.

Wrapped around this energy-efficient body — even the nose — is a dense overcoat of fur. Over most body parts, it consists of two layers: a thin, crinkly inner layer and a thick outer layer of hollow guard hairs filled with air. Together these layers surround the animal with millions of pockets of warm air, trapped both within and between the hairs, making the caribou's coat one of the most amazing thermal insulators in nature.

Although caribou are gifted with a keen sense of smell — they can sniff out energy-rich lichens buried in up to 70 centimetres of snow — they must, like all overwintering mammals, endure occasional periods of food scarcity. When its energy can't be fully replenished by food, a caribou can switch to internal compensation mechanisms that allow it to recycle, resynthesize and store important nutrients from its digestive system. Safety deposits of fat, stored mainly along the back, can be drawn upon during periods of high energy demand, for instance during week-long blizzards when opportunities to feed are virtually non-existent. In the winter, caribou virtually stop growing and lower their basal metabolic rate by as much as 30 percent, further internal adaptations to help shunt limited food energy to all but the most essential body functions, like breathing and keeping warm.

The looks and lifestyle of the snowshoe hare represent another impeccable model of adaptation to winter. This species solves the problem of locomotion in a snow-dominated landscape with its massive hind feet, which, as its common name suggests, act exactly like snowshoes. They are covered in long, stiff hairs that almost double the foot size, permitting nimble passage over the softest, deepest snow in which predators often flounder. More than providing traction, the hare's hind feet provide impressive speed. Down a snow-packed trail, a snowshoe hare can bound along at over 40 kilometres an hour, sometimes covering as much as 3 metres in a single leap.

The hare's ability to doff one coat for another with the changing of seasons gives it an alternative name, the varying hare. Its summer coat is buffy brown, tipped with black. In response to autumn's dying light, its coat turns snow white, beginning with the legs, ears and face and gradually spreading to the flanks and back. By late November, a hare can crouch invisibly against a snowdrift, betraying its presence only with its dark eyes and black-tipped ears.

Snow is an ally to the hare in other ways. Under a heavy snow load, limber birch and alder bushes bend over, bringing their tender growing tips within range of snowshoe hares, which feed on them extensively. When bent, such bushes provide shelter as well as food by forming snow caves where hares take refuge during periods of severe cold, at say −35°C or below. With a roof of insulating snow overhead, much of the hare's body heat is conserved rather than being radiated into the infinite heat sink of the sky. Consistent snow buildup through the winter favours survival of snowshoe hares by keeping a virtually unlimited supply of food within reach. With each snowfall the hare is lifted higher to fresh twigs and virgin bark. When all the snow melts and the bent-over alders and birches spring back into the air, evidence of hare browsing is often positioned metres above the ground. To the uninitiated eye, it may seem that the boreal forest is inhabited by harelike creatures the size of a small bear.

A winter with little snow can be hard on snowshoe hares. If snow remains at a constant thickness or does not knock bushes down to hare height, many animals may die. In such conditions, the hare can sustain itself temporarily by resorting to "refection," an unusual dietary adaptation that involves munching on partially digested droppings and sending them back through its digestive system to soak the most out of every last bit of available nutrition. The hare can stretch limited food resources in this way for several days, but ultimately, if new snow is long in coming, it will be weakened. Those 3-metre bounds will come less easily, increasing the likelihood that it will end up between the jaws of its arch-predator, the lynx.

From the perspective of diet, the body of a lynx consists mainly of reconstituted snowshoe hare. Almost three-quarters of the prey taken by a lynx may be hare, which spread over a year represents a successful kill at least every other day. Like its victims, the lynx relies on flotation to move over the snow, accomplished by its round, well-furred paws, which seem huge in proportion to the rest of its body. Also out of proportion are its long legs and chunky, upraised rump, giving the lynx a generally awkward appearance when standing still. But in motion, the lynx emanates a harmony of smoothly controlled power. Its hunting technique is similar to all members of the cat family: the careful, silent stalk, the crouch, the rush, the kill.

Again like the snowshoe hare, the lynx's chances of winter survival hinge on the condition of snow. Although equipped for travel over snow, the lynx, weighing up to ten times more than a hare, must stalk undetected to within a few bounds of its prey or risk bogging down during the heat of pursuit. Field studies in northern Alberta show that in years when the snow is particularly soft, with little bearing strength, the lynx's attempts to take a hare meet with only 9 percent success, compared to 24 percent when the snow is settled and firm.[92]

So intimate is the ecological link between lynx and snowshoe hare that, about once a decade, their populations peak and crash almost in unison, give or take a year or two. During peak years, hare densities in northern Canada have been estimated at over 6,000 per square kilometre. Understandably, lynx are fat and happy during such years. But the hare population crashes inevitably, usually in winter. After the crash, some lynx may move to other areas, sometimes into communities, in search of alternative prey. Though breeding takes place as usual in late winter, and one to four kittens are born a couple of months later, few make it through the subsequent season of famine. The lynx population takes its crash.

In good times or bad, red squirrels are among the alternative prey for lynx. Marten also eat them, as do ermine, goshawks, horned owls, ravens and occasionally wolves — in fact just about any flesh-eating mammal or bird will pluck a squirrel from a tree if it can. Along with berries and snowshoe hare, the red squirrel constitutes a vital link in many food chains, being one of the most abundant herbivores in the boreal forest. Its continued survival is remarkable, considering that it seems such an easy target for predators. It is among the most active of northern mammals, scurrying about usually in broad daylight. When alarmed, it will more likely chatter loudly and brandish its bright red tail than duck for cover. It is a curious animal and bold, sometimes to the point of being downright insufferable. It pays no heed to my loud threats hurled at it when I discover yet another freshly chewed hole in my cabin wall.

Besides plundering pink insulation from back-country cabins like mine, the red squirrel prepares for winter by hoarding food. Throughout August and September, spruce forests resound with the patter and thump of cones falling to the ground. A red squirrel may spend several hours at the top of one tree, scissoring off its terminal twigs loaded with swelling green cones. It then carries them off to ripen in its winter cache, where several bushels of cones may be stored among the roots of a large tree or under a rock.

Two red squirrels share a joke while dancing on a work table. *(Busse/NWT Archives)*

Cone caches are staunchly defended by squirrels throughout the winter, for without a protected storehouse of food they would soon starve. In a favourite tree near the cache, squirrels perch at regular feeding stations, where they extract energy-rich seeds from the cones. Over generations, the resulting fallout of shucked cone scales accumulates to great depths and can span several metres across the forest floor. Within this mass of scales, which has the consistency and insulating characteristics of shredded styrofoam, squirrels construct needle-lined nest cavities and storage chambers brimming with cones. A labyrinth of tunnels usually runs through these "middens" and connects them to neighbouring caches.

As tempting as this world beneath the snow may seem, red squirrels spend most winter days out in the open, relying mainly on their longer winter coat and the cover of spruce branches for thermal protection. Prolonged exposure to temperatures lower than −30°C is, however, fatal to squirrels. Their size and mass are just too small for them to sacrifice the extra energy needed to stay warm. When temperatures dip below this critical threshold, they take refuge under the snow, snugging themselves into their nests, where they remain inactive for several days. If forcibly brought to the surface before the cold snap breaks, a red squirrel would not survive for more than a few minutes. Only by virtue of this behavioural adaptation to extreme cold does this species exist at all in the Subarctic.

The northern red-backed vole, also called the tundra red-backed vole, is another species that descends below the snow to survive. It does this not just for a few days here and there but for the whole season. Looking like a cross between a lemming and a mouse (from an evolutionary perspective it may well be), the northern red-backed vole displays several morphological adaptations for life in a cold climate: small, well-furred ears, a short tail covered with dense bristly hairs, stubby legs that barely extend beyond its plump, compact body and a winter overcoat of dense silky fur. But like the deer mouse, arctic shrew and other small mammals north of 60°, this vole, weighing about the same as two loonies, is physiologically incapable of processing enough energy to compensate for heat loss to the cold, dry air of taiga winters. Snow offers the only safe place to be.

So effective is deep snow as an insulator that it can be as low as −55°C on the surface, while down on the mossy floor it is just a few degrees below freezing. The most dangerous time of year for small mammals is late September and October, when temperatures can drop quickly before snow cover has built up enough to provide adequate insulation. Many small mammals die during this period of pronounced thermal stress, especially when all that's available for protection against the cold is frozen moss or a dusting of snow.

I once encountered a red-backed vole apparently hanging on through such conditions. It was holed up in a birch log that was rotted out and as hollow as a stovepipe — hardly suitable for a winter's nest. Glued to my tracks, I watched it make a beeline for my boot, sniff it, look me over head to toe, then scramble back to its temporary lair. Was that panic I saw in its eyes? I thought of that animal the next day during a late October snowstorm. I hoped for lots of snow, at least enough to bring its total depth up to 15 or 20 centimetres. Once snow cover reaches this critical depth, heat given off by the earth becomes trapped in the lower snow levels. Seeking out this warmth, red-backed voles vanish from the surface, safely retiring to a maze of snow tunnels and burrows for the rest of the winter.

Living mostly in semi-darkness or, at best, a soft bluish light, the vole attunes to its snowy underworld primarily through an acute sense of touch, achieved through long, sensitive whiskers around the face and a bracelet of special short hairs around the wrist.

A keen sense of smell and its ability to detect subtle vibrations, like the footfall of an approaching fox, round out the vole's main sensory gear. Thus equipped, the vole can manoeuvre effortlessly through its complex system of tunnels, stopping at a food cache for a meal of berries or buds, checking one of its many sentry posts at the edge of its thousand-square-metre territory or visiting a nest chamber to groom its fur and have a quick nap.

So goes the life of red-backed voles in winter. Their world is warm, moist and remarkably stable compared to the wide weather fluctuations up top. If the snow cover is especially thick, they may be virtually immune to detection by foxes or owls, their most frequent predators. However, a thick blanket of snow is not without inherent dangers. The same warmth that sustains voles allows a limited amount of bacterial action to continue in the upper soil, even in mid-winter. This results in a gradual buildup of carbon dioxide under the snow, particularly when it is trapped by a series of crusty layers above. By late winter it can reach potentially lethal levels. Fortunately, voles are able to cue in to rising carbon dioxide levels and respond by building ventilator shafts to the surface. Unfortunately, for the voles at least, these shafts are an open invitation to predators.

Ice is the winter ally of the beaver. In a pond of its own creation, this animal carries out many of its activities — foraging, evading predators, even mating — under a metre or more of ice. Stuck into the pond's muddy bottom is a winter's supply of trimmed aspen, birch and willow sticks stashed there by beavers in the fall. On a visit to its feed bed, a beaver may choose to eat on the spot, gnawing off a bark meal with its long, chisel-like teeth. Swallowing too much water while feeding is not a problem. Special flaps of skin behind the teeth keep water out. Nor is oxygen a problem. A beaver's heart beats much slower than that of other mammals of equivalent size. It can easily stay underwater for fifteen minutes. If, after this time, the beaver's hunger is not satisfied, it may suck on a few air bubbles trapped under the ice, some of them formed from its own exhalations. Or it may drag a stick or two up an underwater tunnel to its lodge.

The beaver lodge is a natural masterpiece of construction, complete with a raised sleeping platform covered in insulating shreds of bark, two or more plunge holes opening to the main entrance and auxiliary escape tunnels, an air vent at the thin apex of the lodge, and inner walls covered in a frozen mud plaster as thick and firm as adobe. Insulated from extreme cold and virtually impregnable to potential predators such as foxes, wolves or lynx, the lodge provides a safe winter fortress in which the routines of a beaver's family life unfold.

When not tending the needs of young kits, eating or sleeping, adults spend much time preening their rich, dense fur — that luxurious resource that set the direction of northern Canada's history for two hundred years. At the base of the beaver's paddlelike

Though invisible for most of the season, beavers are active all winter long. While locked under snow and ice for up to six months, they rely on a well-fortified lodge (left), a well-stocked food pile (right of centre), and a dam (far right) high and strong enough to hold back at least two metres of water.

tail are glands that discharge waterproofing oil onto the fur. Using its forepaws, the beaver first brushes this oil over its entire coat. Then, with a specially adapted set of split nails on its hind paws, it carefully combs the oil deeply into its fur. Regular maintenance of its sleek oiled coat allows the beaver to continue an aquatic lifestyle through the winter, even in the coldest water.

Except when lodged beside a large river, there would be no water to swim in if it were not for the dam — and not just any dam, but one big enough to impound water to a depth of at least a couple of metres. Anything less than this and the pond could freeze nearly to the bottom, imprisoning the beaver in its lodge or restricting access to its food piles. Dam building and repair work go on throughout most of the summer and fall as a cooperative project for a family of around four beavers. Some members fell and trim trees, while others haul them away and position them on the dam. Another concentrates on packing the dam with mud and stones. The finished dam curves into the woods along both banks and slopes gently into the pond on its upstream side. It is built in the strongest configuration possible for holding back water. A team of human engineers equipped with nothing more than axes probably could not improve on its time-tested design.

The inborn trigger that stimulates a beaver to build a dam is the sound of running water. It so happens that the best place to build a dam — the narrowest points of a river or creek — are also the noisiest. When a beaver moves into a new area, this is where it starts chopping. To demonstrate the connection between babbling water and the beaver's urge to dam it, an animal behaviour scientist devised a simple experiment. In a laboratory setting, a group of seasoned adults was induced to build several dams at pre-selected points inside a large tank of dead-calm water. They were responding instinctively to the sound of fast-flowing water — coming out of a tape recorder.

In the wild, beavers may build several secondary dams downstream from the main dam, creating a steplike series of ponds. Collectively, these dams bring the beaver's main food and building material, aspen trees, closer to the water, an arrangement that minimizes the risk of predation on land and reduces the beaver's workload. Beaver dams also stabilize water flow, reduce streamside erosion and provide productive wetland habitats for many forms of wildlife besides beavers. Puddle ducks nest in the marshy shallows. Muskrats feed along the shore. Insects lay eggs in the water, which develop into larvae eaten by wood frogs and many species of fish. Yellow pond lilies and other aquatic plants take root in the rich muck that accumulates at the bottom.

For beavers, the yellow pond lily is a preferred food for much of the summer. Why eat trees when such succulent, protein-rich food can be eaten in the safety of the pond? But as autumn approaches and the air begins to cool, beavers turn their attention almost exclusively to felling, trimming and stockpiling trees. This time of year, the beaver family presents a picture of diligent activity as it works together towards a single end — to store food and fortify its shelter in preparation for the deep freeze of winter.

Once winter locks in, the beaver's world appears dormant and still, with the lodge, the pond, the dam, the rocky shore all blending together under a uniform blanket of snow. Unseen below this protective cloak, beavers maintain an active pace of life. As snug and secure as their winter world may seem, hibernation is out of the question. As long as they make regular forays out to their under-ice food piles, beavers can enjoy the benefits of a lodge warmed by central heating — supplied by their own bodies. But for a family of beavers to curl up into a ball hoping to wait out winter by conserving metabolic energy would be suicidal. After a few days of inactivity, their metabolic furnaces would shut down, resulting in a frozen huddle of beavers in a frozen lodge. Beavers must feed throughout the winter, since they are not metabolically equipped for hibernation.

Along with food caching and thick fur, hibernation is another one of nature's ways of carrying an animal through the acute energy stress and food shortages of winter. When hibernating, an animal's metabolism, heartbeat and respiration drop to a fraction of normal. Like a cold-blooded reptile, its body temperature plunges to match that of its surrounding environment, the hibernating den. If the den temperature dips below freezing, the animal is awakened by a violent fit of shivering, which raises its body temperature before ice crystals can form in its tissues. The process of waking and rewarming takes two to three hours and is itself a tremendous physiological drain.

As it turns out, except for the odd arctic ground squirrel along the region's northernmost fringe, there are no true hibernators in the taiga shield.[93] Contrary to folklore, black bears are no exception to this rule. Strictly speaking, they do not hibernate. In early October, they do retire to sleeping dens, usually in a rock crevice, under a wind-thrown log or at the base of a large spruce with low, sweeping branches. And, like hibernators, their heartbeat and respiration drop dramatically. But their body temperature drops only a few degrees, from a normal 38°C to a low of 31°C — nowhere near freezing. For most of the winter, a bear remains immobile in this state of torpor, a kind of lethargic dormancy. The only sign of life might be a thin column of vapour rising from its snow-covered den on crisp, still days. Unlike hibernators, a bear can be aroused more easily from its winter sleep. The scientific literature tells of bears woken up in a matter of moments by a few hand claps, shouts or a good poke.

For a slumbering bear, the official end of winter may come sometime in April with a sudden cave-in of soggy snow or a flash flood of meltwater soaking the den floor. Where the snow has settled thick and firm, a bear may have to bore a tunnel to the surface. They emerge ravenous. With snow still on the ground, the pickings are slim. Its first meal may consist of spruce needles, tree buds or the discarded hind leg of a snowshoe hare. It may tear apart a rotten log in search of a handful of ants, grubs or beetles. If it can get it, cardboard or rags will do. At this time of year, a bear out at my cabin ate some rubber hose and sampled the plywood from my outhouse door.

Black bears are not fussy eaters. They are omnivores, a term derived from the Latin words *vorax*, to devour, and *omnis*, all. They'll eat just about anything. During the summer and autumn, over 75 percent of a bear's diet consists of plant matter, with berries being the main attraction, as testified by its purplish scats chock full of seeds. The balance of its diet is made up of grasses, roots, insects, small mammals, birds' eggs and carrion — the remains of previously killed animals. Being opportunistic feeders, black bears will not pass up the chance to seize a moose calf or other vulnerable game.

While out for a quiet evening paddle on River Lake, I was once startled by a series of tremendous splashes along the shore. The hair on the back of my neck stood instantly on end as I caught sight of a black bear crashing through the shallow water towards my canoe. Fortunately, I was downwind from the bear and its attention was fixed on a retreating northern pike, not on me. Absorbed in the chase, it bounded past me like a 100-kilogram puppy, missing my canoe by only a few metres. Like their Kodiak cousins stalking salmon along the Alaskan coast, black bears will sometimes spend hours fishing like this along shallow lakeshores or slow-moving streams.

The black bear's wide-spectrum diet is one of the main reasons for its success as a species. It ranges from the deserts of northern Mexico to the frigid Beaufort coast. In the taiga shield, this flexible feeding strategy is crucial to its survival, guaranteeing that fat reserves will be sufficiently built up by autumn to carry it through the long period of winter dormancy.

Counting all the mice, voles and shrews, all the weasels, canids and other carnivores, all the ungulates, plus all the tundra dwellers, like the grizzly and arctic fox, that make regular sojourns into the trees, there are about fifty species of mammals inhabiting the taiga shield. For many of them, the deep snow, cold temperatures or scanty food supply of winter are among the most powerful forces influencing their life cycles. Shield country mammals are creatures of winter. How they adapt to it is central to the drama and wonder of their lives.

12

Birds: Poised at the Crossroads

In the remnant jungles of India, there are watchtowers where tourists go to experience exotic wildlife close up. From these elevated wooden platforms perched between trees, you might, on a good night, see a troupe of elephants, a wild boar or maybe even a tiger. On this May morning north of 60°, my watchtower is a raised hummock carpeted with sweet-smelling Labrador tea. Leaning back against a black spruce, I have come to sit for a while beside this still lake to witness the explosion of spring.

Throughout the great Canadian muskeg there are tens of thousands of these small peat-bottomed lakes and ponds, most of them nameless. Their shallow depth and heat-absorbing peat cause them to break up long before the bigger, deeper lakes, which can remain locked under ice until well into June.

Winter is dealt a death blow with the early thawing of these smaller water bodies. Acting like biological magnets, these scattered patches of shimmering dark-blue water are irresistible to the surging tide of migrating ducks and other waterbirds, which are what I have come to see.

I peer, amazed, through my spotting scope at a tiny pan of rotting ice covered in loafing ducks: northern pintail, American wigeon, green-winged teal, northern shoveler and lesser scaup. Crowded bill to bill, they look a bit travel weary, having journeyed from the Pacific or Atlantic tidewaters, the coastal marshes ringing the Gulf of Mexico or even farther. A flotilla of surf scoters, buffleheads and canvasbacks glides into view. The collective whistling, peeping and yodelling of ducks is music to my ears, long starved for the sounds of spring.

Suddenly, above my head, knife edges carve the air. With landing gear down, wings arched and unflapping, six old squaw ducks plummet towards the lake, circle once, then set down in each others' spray. On their way to arctic nesting grounds, they have chosen this safe haven to rest, feed and engage in zealous courtship.

Arctic and subarctic ducks mingle in the hundreds. Swans and geese often land here, as do numerous shorebirds, gulls and terns. The music of songbirds pours out from the lake's shrubby margins and neighbouring forests: the spiral fluting of the Swainson's thrush, the gurgling "konk-la-reeee" of the red-winged blackbird, the "sweet, sweet, Canada, Canada, Canada" proclamation of the white-throated sparrow. Sweet it is here at Niven Lake, a five-minute walk from my Yellowknife home.

The mid-May arrival of vast numbers of birds is one of my favourite events in the annual cycle of the seasons. At this time of year, I can sit on my deck enjoying genuine warmth in the sun's rays and listen to the hollow rattle of sandhill cranes far overhead or watch an endless aerial parade of gulls fly past in search of open water. Gripped by spring fever, I sleep restlessly at night, often woken by the haunting calls of north-bound white-fronted geese, Canada geese or tundra swans. While tackling a sink full of dishes, I am startled by a low-flying bald eagle just metres from my kitchen window. A twittering flock of Lapland longspurs lands on the railing of my fence and I am awestruck.

Outside of town, along the north shore of Great Slave Lake, the spectacle of migrating birds attains its grandest proportions. Flowing south into the lake are over twenty creeks and rivers, the lower ends of which are among the first bodies of water to open in the spring. Before finally merging with the waters of Great Slave, many of them drain into large, open wetlands and shallow bays. Along the north shore, there are thousands of rocky islands, the edges of which absorb much more heat than the surrounding ice. Rings of open water form around each island, gradually expanding and merging together. From the Frank Channel in the west to the Hearne Channel in the east, there are, by mid-May, at least a thousand square kilometres of open water available along the north shore. Each spring, staging waterbirds gather here in the tens of thousands.

"Their numbers are outrageous!" according to Jacques Sirois, a Yellowknife ornithologist working with the federal government's Canadian Wildlife Service. Sirois spends a good part of May in the air, cruising thirty metres above Francois Bay, Drybones Bay, the lower Beaulieu and Yellowknife rivers and other staging hot spots along the north shore. "Although most of Great Slave Lake is still frozen over, there's a lot of shallow open water down there and the ducks, loons and grebes know it. At first the birds seem widely dispersed, but once you start counting, it's spectacular."

On a typical survey flight during the peak of spring migration, Sirois might record over 2,000 tundra swans, 20,000 Canada geese and 15,000 scaup ducks. Seeing in a single

All feet and no feathers, this young bald eagle strikes a defiant pose. *(Busse/NWT Archives)*

pass 1,000 mallards, northern pintails and American wigeons is routine for Sirois. On a particularly fruitful day towards the end of May, he once logged almost 50,000 waterbirds representing 26 species, 4,000 of them from the Yellowknife River alone. Sirois estimates that some springs, particularly around late May, as many as 100,000 migrating waterbirds may stage daily in the open waters and wetlands along the Great Slave's north shore. That's ten times the established standard for recognizing waterfowl concentrations of global significance.

North America's migration corridors for waterfowl have been mapped out in great detail over the past fifty years using various forms of technology and scientific wit. Millions of flight records from radar observations, radio telemetry and bird band recoveries have helped locate these corridors. A less sophisticated but perhaps more exciting technique is the early practice of attaching a small penlight battery and coloured bulb to a duck's leg and tracking its flight path at night by car — at least for a few kilometres. Collectively, these observations show a spaghetti-like pattern of migration corridors concentrated along a northwest-southeast axis stretching from the Gulf of Mexico to the Mackenzie Delta, one of the world's most important waterfowl breeding grounds. Many waterfowl that stage along the north shore of Great Slave Lake are on their way to this delta. Via Lake Athabasca, the Slave River Delta, Great Slave Lake and Great Bear Lake, they follow a migration corridor that exactly parallels the low northwestern edge of the Canadian Shield.

The shield edge offers more than just plenty of open water for migrating waterfowl. Along a narrow zone of taiga shield, from Lake Athabasca to Great Bear Lake, the shallow bays, small ponds and other wetlands show a host of striking differences from those on the shield's interior to the east. In general, water bodies along the shield edge are smaller and shallower, they tend to have muck rather than sand or gravel bottoms, their shorelines are dominated by trees, shrubs or sedge mats rather than bedrock, their waters are warmer and less acidic, and they tend to have more abundant aquatic plants, which are of particular interest to waterfowl, whether migrating or breeding. In short, the wetlands of this narrow zone are much more fertile. Why?

The answer lies in this zone's unique glacial history. In most shield ponds, bottom sediments — the foundation of a pond's fertility — have built up slowly since the last glaciation, when they were scratched from the relatively acidic and unyielding shield bedrock. It's a different story along the northwestern edge of the shield. It turns out that this area corresponds remarkably well with the eastern shoreline of Glacial Lake McConnell, which inundated this area from about 11,000 to 8,000 years ago.

This lake received torrents of thick sediments washed from the melting glacier and the adjacent landscape, just freshly exposed. These included fine silts and clays from the softer, more alkaline sedimentary rocks west of the shield edge. It is these lacustrine (or lake bed) sediments, particularly near the former shoreline, that have enriched the northwestern edge of the taiga shield and made its wetlands so important to waterfowl. The effect of this enrichment is so pronounced that the density of breeding ducks along the shield edge, about twenty pairs per square kilometre, is at least twice that of wetlands just a little to the east.

Among the marvellous mix of duck species that make the taiga shield their summer home, the first to arrive are the dabblers, or so-called "puddle ducks." Feeding in shallow waters, these birds skim the surface with their beaks or tip up, tail skyward and wiggling, to reach submerged plants or animals living on the bottom. When spooked by a potential predator, hunter or nosy naturalist, a dabbler takes off in an explosion of spray, lifting itself free of the water into full flight with just a few powerful thrusts of its broad wings.

Smallest of the dabblers is the green-winged teal, often called the greenwing because of its brilliant green wing patches visible in flight. During aerial courtship chases, greenwings dart about like shorebirds, flashing their white bellies as they twist and turn in the air. These are true puddle ducks, nesting sometimes beside the smallest of water bodies — a flooded roadside ditch or a barely trickling creek. They are reclusive birds, concealing their nests in the thickest of brush and rarely venturing beyond the cover of shoreline reeds and grasses. But greenwings have a strong presence in this region, making themselves plainly known when sunlight ricochets off the male's splendid green head band or when his piercing froglike peeps echo through the boreal woods.

The American wigeon is probably the most common dabbling duck in the Northwest Territories. Over half a million breed annually on the taiga shield and adjacent Mackenzie Valley. A distinctive field mark of the wigeon is the shining white crown of the male, giving this species another common name: the baldpate. On the water, wigeons display a graceful poise, riding high and pivoting from side to side as they lightly peck at the surface for food. During the breeding season, males frequently issue three mellow whistled notes — "whee whee whew" — while the females announce their presence with a hoarse, rather stereotypical "quack."

Outstripping the wigeon not in numbers but in range is the northern pintail. The same race of pintails that nests along the shores of Lake Athabasca is found well above the Arctic Circle along the Beaufort coast. It also breeds across northern Siberia, Scandinavia and Greenland. In winter, the pintail flies as far south as Columbia, the West Indies and tropical islands in the south Pacific, making it the farthest-ranging species of waterfowl on earth. This claim to fame couldn't go to a more attractive duck. The pintail's trim appearance and swift flight have given it the nickname "greyhound of the air." The male's breeding plumage is unmistakable. The snowy whiteness of his belly and breast curls up in a half moon along his delicate neck. His silver-blue body tapers to a slender tail made especially long by two needle-pointed feathers, which give the bird its conventional name. The Dogrib people of this region have given this bird a different name, *aawa*, derived from the wheezy double-toned whistle uttered by the courting male.

By the time the major influx of dabblers is over — usually mid-May — the "divers" start pouring in. Searching mainly for aquatic insects and other invertebrates, diving ducks propel themselves deep underwater with wide paddlelike feet positioned far back along their bodies. Because their wing area is relatively small in proportion to their body weight, divers can't leap suddenly into the air like dabblers. Instead, they nose into the wind like a floatplane and literally run out of the water.

One species of diver once made the headlines in a Yellowknife newspaper: "Rare duck nests in YK family's backyard birdhouse."[94] What was thought to be a woodpecker of some kind turned out to be a not so rare bufflehead, a diminutive black and white duck that happens to nest in trees (and also birdhouses, apparently). The breeding distribution of this boreal bird is limited only by the availability of vacant tree cavities, usually hammered out by a flicker, within waddling distance of water. The flightless ducklings must cover this distance on foot. Obeying the quacked urgings of their mother, they leap from their tree trunk nest and free fall up to fifteen metres to the ground. Once they stop bouncing, the ducklings are led to water, where they can swim and dive for food without instruction. Unless snatched from the surface by a hungry pike or herring gull, by late August the young buffleheads will be winding and weaving centimetres above the water in the close-knit flocks typical of this species.

The red-breasted merganser takes a straighter course through the air, its rapid, shallow wing beats pumping it along at speeds up to 130 kilometres an hour in level flight. The male sports a green-black crest, which appears swept back by the wind of its swift passage. A handsome rust-coloured breast band complements this bird's rakish appearance and clearly distinguishes it from the whiter common merganser. On the water both species typically ride low, with their tails awash and water almost up to their wing tips. Underwater, they move with all the speed and agility needed to catch fish, their main source of food.

A successful underwater chase ends as the merganser clamps its serrated beak onto a fish (hence the scientific name for a red-breasted merganser, *Mergus serrator*). But catching its prey is sometimes only half the battle. I once saw a male merganser surface with a wriggling fish in its beak. Pandemonium broke loose among the flock as it set upon him in a struggle to snatch the fish away before he could swallow it. While being pursued, the fleeing merganser managed to flip the fish around in its beak and, as it must, swallow it head down. The rest of the flock all dove instantly, as if to sulk, leaving the victorious bird drifting quietly on the surface, occasionally sipping water to wash down his hard-won meal.

"Black at both ends and white in the middle" describes the taiga shield's most common diving duck, the scaup, so named for the loud "skahp" note it emits when alarmed. Two species come here to breed in the spring, the lesser scaup and the greater scaup. Telling them apart is not easy. On the water, you can do it *if* they are fairly close and *if* the light is good; the less common greater scaup is a slightly larger bird with a rounder, green-tinged head, in contrast to the blacker, more domed head of the lesser scaup. On the wing, the key is a broad white wing stripe, which is relatively longer in the greater scaup. Together, they migrate north in huge flocks and nest here in numbers unmatched anywhere else on earth. Scaups reach their highest breeding densities, about 16 birds per square kilometre, along the fertile edge of the shield.

Over 40,000 scaups breed along the pond-dotted Slave River Valley and on the countless islands and inlets of Great Slave Lake's north arm. Thousands more nest on the many ponds, lakes and rivers of the interior taiga shield. Whether breeding beside a still bog or an exposed rocky shore, most male scaups leave the females and young behind and flock to the big, deep lakes by early August. Here they once again gather in the thousands, joined usually by scoters, mergansers and other diving ducks plus several kinds of gulls.

This is the moulting time, when waterfowl shed their breeding plumage in exchange for an entirely new set of feathers. Flightless for about three weeks, the diving ducks drift far out from shore, where they can evade predators while still able to dive deep for food to meet the considerable energy demands of making feathers. Next to the spring migration in May, these huge, bobbing "rafts" of moulting ducks in August are one of the region's most amazing bird spectacles.

The greater scaup is one of several birds whose summer range corresponds remark-ably well with the northern taiga fringe of the boreal forest. The gray-cheeked thrush, northern shrike and Harris's sparrow are other species that are almost exclusively taiga breeders. These species are not found in the dense coniferous forests to the south. Nor do they venture much beyond the northern limit of trees. They are truly taiga species, preferring to live on the edge or "ecotone" between two of the world's largest biomes, the boreal forest and the arctic tundra.

Being a zone of transition between two very different biomes, the taiga shield offers a relatively wide variety of habitats for birds. Here, boreal lakes, wetlands and forests are interwoven with open shrublands and sedge mats more typical of the Arctic. The consequent overlap of northern and southern birds gives this region a special richness in bird diversity.

At or near the *southern* limit of their summer range are such common "arctic" birds as the red-throated loon, arctic tern, red-necked phalarope, parasitic jaeger, semipal-mated plover and common redpoll. A host of water birds and boreal forest species reach the *northern* limit of their range on the taiga shield, including the Caspian tern, common tern, California gull, bufflehead, hermit thrush, orange-crowned warbler, Tennessee warbler, white-throated sparrow and chipping sparrow. And the city of Yellowknife boasts the continent's most northerly population of a bird that originally hailed from north Africa and the Middle East: the house sparrow.

Superimposed on this mix of birds at the northern and southern limits of their breeding range is the mingling of eastern and western birds. Typically western species include the white-winged scoter, canvasback and bohemian waxwing. Birds found here that are typical easterners include the eastern phoebe plus several warblers: the palm warbler, Magnolia warbler and the black and white warbler.

These lists are not comprehensive. They may change with the weather. For instance, with a dry year on the prairies, ruddy ducks, redheads, American coots and even yellow-headed blackbirds may venture north in search of well-watered breeding habitat. A poor spruce cone crop in northern Alberta may bring north large flocks of white-winged crossbills, a species usually uncommon in this region. A late spring on the barren lands may increase the number of Canada geese or oldsquaw choosing to nest here rather than above the tree line. A prolonged nor'wester from Alaska may blow in a few northern wheatears or long-tailed jaegers, much to the delight of local birders.

To the keen observer, each year brings subtle variations in the abundance and variety of this region's bird life. But the basic cycle of events is as inevitable as winter: the sudden deluge of spring migrants, the loud and lovely staking of territorial claims, the feeding and fledging of the young, the mid-summer rafts of moulting ducks and mixed flocks of bug-hunting warblers. And then, for most birds, comes the time to prepare for the reverse journey home to their wintering grounds.

As much as those of us who live here may take pride in "our" ducks, thrushes, warblers and other breeding birds, many are best thought of as residents of the tropics, since they may spend over two-thirds of the year in steaming jungles or seaside mangrove swamps far to the south. Journeys of three or four thousand kilometres are not unusual. White-crowned sparrows winter as far south as the dry chaparral of northern Mexico. Some barn swallows spend most of the year chasing insects through the Brazilian rain forests. The undisputed champion of long-distance migration is the arctic tern, flying south in the fall to such destinations as the southern tip of Argentina, or beyond to the ice-bound edge of Antarctica — a round trip journey of around 35,000 kilometres.

For many birds such as swallows, flycatchers and warblers, it's shield country's ubiquitous bugs that make this such an attractive breeding area and it's bugs that fuel the journey back south. Long-distance migrants must lay down considerable stores of fat before leaving. The shorter days and cooler temperatures of August stimulate intensive feeding behaviour in some birds, which over the course of just a few weeks may double their body weight. For instance, the tiny blackpoll warbler averages a mere 12 grams in early summer but may weigh 20 to 25 grams when it sets off on migration in late August.

Many ducks that summer in shield country need nothing more than open water and the food below to make it through the winter. Most of these are diving ducks, including mergansers, goldeneyes and the white-winged and surf scoters. Instead of expending tremendous amounts of energy on a long flight south, these birds migrate relatively short distances to open seas along the Pacific and Atlantic coasts. At least, most of them do.

During the 1990 Christmas bird count in Yellowknife, I was amazed to discover a common goldeneye and lesser scaup loafing on the ice beside a patch of open water warmed by the local power plant. With the temperature at −30°C, it was clear that the cold was not a mortal hazard to these birds — ducks are remarkably well insulated creatures, and not just diving ducks. From the interior of Alaska comes a report of several mallards dabbling beside a lively mountain stream at −45°C.

"Partial migrants" are birds whose flight south for the winter may involve little more than a short hop from the tundra to the taiga. Most warblers must fly great distances to gain access to a winter's supply of insects. Diving ducks must fly across half a continent before finding water that's open all year. For redpolls that breed on the tundra, their greatest need in winter is a secure source of small, protein-rich seeds. To find them they need travel no farther than the nearest birch tree.

There are two kinds of redpolls, the common and the hoary. Both are small, greyish-brown birds with a bright red cap tipped low over the forehead. The males of both species have a black goatee-like patch on their chin and a pink blush on their breast. The hoary redpoll is generally lighter in colour, having a more "frosted" look to it — hence the name — and it has an unstreaked whitish rump, the best clue for telling redpolls apart. Among ornithologists, a debate has simmered for years over whether these birds are in fact one or two species.

Oblivious to their questionable identity, redpolls form mixed flocks that cruise the taiga forests from October to April. Descending upon a birch tree with a rattling "chet-chet-chet-chet," a flock will extract the seeds from its tassel-like cones, then hustle away suddenly, leaving behind a shower of tiny scales on the snow.

Like redpolls, the willow ptarmigan and rock ptarmigan are primarily winter visitors to the taiga shield, flying north to the arctic tundra for the breeding season. These species are also tricky to tell apart. Both turn pure white in winter except for their jet black tail feathers, which show best in flight. Both have densely feathered feet, which, like the rear thumpers of a snowshoe hare, enable them to float over the softest snow. In winter, the definitive difference between the males of these two species is a sliver of black extending from the bill to the eye; if it's there it's a rock, if it's not it's a willow.

Like giant snowballs, ptarmigan dangle from lakeside willows while snipping off tender buds with their heavy chickenlike beaks. Snipping their way through a patch of willows, they make short, fluttering vaults from bush to bush. If alarmed — and it often takes a lot to alarm a ptarmigan — they will rocket into the air en masse and fly a straight and sure course for cover. To conserve precious body heat between meals, ptarmigan regularly plunge headlong into the snow to seek an insulated roost below the surface.

Notwithstanding a few errant ducks, snow buntings or the odd gyrfalcon, bird watchers in shield country generally can expect to see about a dozen species at best during the annual Christmas bird count. Redpolls and ptarmigan are among the regulars. Also included is a group of "hangers on" — birds that don't migrate at all — such as the spruce grouse, boreal chickadee, gray jay, common raven and house sparrow. Through their evolution, these birds have traded off the tremendous energy expense and in-transit hazards of migration for the risks associated with the long subarctic winter. Thanks to various physical or behaviourial adaptations, they all have solved the number-one problem that forces so many other birds to leave: getting food in a cold and dark landscape covered in snow.

The house sparrow's ability to overwinter in this region is largely dependent on human benevolence. If Yellowknife were to become a frontier ghost town, the city's 300 or so house sparrows would have to take up residence elsewhere, like the people who once sustained them through their backyard bird feeders.

The gray jay is a more typically northern winter resident, so typical that it has been given more common names than any other boreal bird. Until a few years ago it was officially the Canada jay. Among old-timers, it was (and still is) the whisky jack. For those who spend a lot of time in bush camps, it is the camp robber, perhaps a title well deserved. But who could hold a grudge when weighing this bird's petty crimes against its abounding charm?

The gray jay can be a conspicuously noisy bird, with some notes harshly whistled, others softly warbled or cooed. Or it can blend itself unobtrusively into the deep fabric of boreal stillness. Its loose, fluffy plumage and gliding flight allow it to swoop out of nowhere as silently as an owl. While wandering the secret ski trails behind my cabin, I once was alerted to the presence of a gray jay just above my head by only the faint click of claws meeting a branch. As dark, inquisitive eyes looked down at me, I got the distinct impression that the next move was mine. The jay watched fearlessly as I pulled a bag of sunflower seeds from my pack and spread a few on my outstretched hand. As if this scene had been rehearsed many times over, it jumped unhesitatingly onto my hand, downed a couple of seeds, then carried a few off to a large spruce to cram them under some loose bark.

Without this habit of food hoarding, the gray jay could not survive winters on the taiga shield. Hoarding goes on mostly in summer — hoarding of berries, insects, mushrooms and meat — with hidden stores being drawn upon throughout the winter. This survival strategy demands an unusual level of intelligence for a bird. Not only must they hide hundreds of bits of food from would-be thieves, but they must remember where the heck they put them.

A gray jay perches nonchalantly on the camera of Yellowknife naturalist Bill McDonald. *(Busse/NWT Archives)*

The gray jay's apparent success at hoarding confers another benefit to its way of life. With its winter food supply ensured, it can begin nesting activities as early as March. Before some migratory birds have even arrived, the year's crop of young jays is flying about, already learning how to hoard food for the next winter.

My daily link to the bird world is the raven. All I have to do is take a few moments to scan the sky for its black silhouette or listen for its hollow croak and one is there, even on the most frigid days of winter. To see a raven sculling nonchalantly through the ice fog at 40° below seems incongruous, to the point of being brash — especially when most other creatures, except those in parkas, are lying low to save precious energy. The raven makes itself at home throughout the year in places as far north as Ellesmere Island and the northern coast of Greenland, where it coexists with only a handful of other species hearty enough to survive the planet's most extreme arctic conditions. How does it do it?

Some features of the raven's body help it withstand the cold: a heat exchange blood system in its legs to keep them from freezing solid, an ability to fluff up its feathers to enhance their insulating effect, feathered nostrils to conserve heat and moisture, a thick, powerful bill to tear into frozen carrion, plus an extremely high metabolic rate — a real furnace — one of the main reasons for its voracious appetite.

But none of these features is unique to the raven. And none helps this species adapt to the food limitations of winter. The raven does not have a sharp, tapering beak to extract seeds from a spruce cone or snip buds from a willow. It does not have acute stereoscopic vision, like most birds of prey, to help pinpoint its quarry from high in the air. It does not have a particularly sensitive nose to scavenge for carrion. It does hoard food occasionally, but not nearly to the same extent as its close cousin the gray jay. To make it through winter in this part of the world, the raven relies mostly on its brain.

Wherever the raven is found, it has figured prominently in the legends and lore of aboriginal people — the Yaqui Indians of northern Mexico, the Haida along the British Columbia coast, the Inuit of the Arctic Archipelago and the Dene of shield country. Some of their stories depict the raven as a saviour or sage, others as a trickster and a cheat. In all of them, this bird is set apart from the rest of the animal kingdom by its cleverness. This reputation comes to the raven by its well-demonstrated ability to learn quickly from experience, to respond flexibly to changing circumstances and to communicate information to its own kind and, say some scientists and shamans alike, to humans.

The raven is an immensely adaptable bird, able to use its intelligence to find food in winter through scavenging, predation and sheer opportunism. Like gulls following a fishing boat out to sea, ravens will follow a pack of wolves or a bear in expectation of sharing spoils from some future kill. Ravens are often attracted to human hunters by the sound of shooting, having learned that gunshots mean death and a likely carrion meal. As an accomplished predator itself, the raven relies on boldness and strength to capture its prey, which may include voles, weasels and even ptarmigan.

Around human settlements, the raven's winter is a cycle of progressive dinners with one main course: garbage. Besides the town dump, garbage cans are a favourite target for this wily bird. I have watched a couple of ravens land on the edge of a can, then rock it back and forth until the contents spill on the ground for looting. At sunset, hundreds of squawking, well-fed ravens will congregate at their nocturnal roosts to compare notes on

The immensely adaptable raven. (*John Poirier*)

the day's exploits. On one December evening in 1991, biologists counted over 800 ravens perched on rooftops and railings within two city blocks in downtown Yellowknife. In spite of the fact that this phenomenon of roosting ravens may be a by-product of our own garbage, the constant commotion of their dark, shuffling bodies and cacophony of their calls make it one of the region's most outstanding winter spectacles.

Like the gray jay, the raven's success at overwintering allows it to begin reproductive activities much earlier than most birds. Serious courting begins in February and March, a time when the flight displays of ravens are at their acrobatic best. With unbridled passion, a pair of birds will play for hours on the wing, tumbling, twisting, somersaulting — even flying upside down for a few seconds. They may climb steeply together, almost out of sight, then plummet back to earth with folded wings, playing aerial tag on the way down. At this time of year ravens are also most verbose. They have a repertoire of about thirty sounds to draw upon, the classics being a throaty "crrruk" and a mocking "who-haw."

As the spell of winter breaks and nesting time approaches, the usually gregarious raven retreats into relative seclusion and silence, its voice overshadowed for a while by the honking of swans and geese overhead and the laughter of returning gulls in search of open water.

13

Insects: Sky Sharks and Bulldogs

Strange things are done under the midnight sun in the name of science. Somewhere out on the barren lands north of Yellowknife, a team of dedicated entomologists — bug scientists — once stripped down to their underwear and stood stock-still for several minutes, humbly offering their blood to a choking swarm of mosquitoes. Meanwhile, a trusted assistant (presumably clothed) scurried from body to body recording the number of mosquitoes feasting on the exposed human flesh.

The final tally: as many as 9,000 bites per minute. At this rate, half the blood supply of an unprotected human could be pumped into the bellies of mosquitoes within two hours — easily enough to cause death. The entomologists reported no documented cases of death by "exsanguination": the sucking of blood. Pursued by bugs, a person running half-crazed off a cliff is a more believable and equally regrettable way to go.

Why this frenzied lust for blood? For at least 150 million years, mosquitoes have been sucking flower nectar to meet their daily energy needs. They still do it, the males exclusively. In some species, the female too can carry out all her life functions, even nourish her developing eggs, without consuming a drop of blood. Next time you inhale a mouthful of mosquitoes, take a moment to reflect on their taste. A trace of sweetness? Probably. This reflects their continued dependence on nectar. But blood is much richer than nectar, particularly in proteins, and can increase egg production a hundredfold.

Getting a blood meal is risky business, especially for a tiny, fragile insect approaching an unwilling donor many times its size. Consider the outcome if you were to approach an elephant with a syringe. Hence many species of mosquitoes resort to various kinds of stealth behaviour, such as attacking at night, when their slumbering hosts are unable to strike back. Others rely on safety in numbers — big numbers, the kind that only physicists can understand — so that at least a few make it through the front lines of battle without getting squashed.

A hungry female sniffs out her unsuspecting prey through various cues, such as carbon dioxide, heat and certain chemicals given off by our skin. Having found a host,

Mutant mosquito buzzes caribou herd while slinging a cat away for a blood meal. Photo fable created by Yellowknife photographer Henry Busse during the long winter of 1953. *(Busse /NWT Archives)*

she sets to work with her remarkable mouth sawing her way into the skin. Like an oil rig, she can probe in several directions looking for buried treasure. Finding a vein or artery is a bonus. Usually she relies on blood from small pools created by all her sawing back and forth through capillary-laden tissue.

Once she is locked onto a blood supply, the sucking begins, powered by two head-mounted pumps. To keep the blood from clotting as it is pumped, she first injects saliva into her victim. Proteins in her saliva cause that infernal itching and swelling, lasting mementos of a mosquito's visit. No wonder one common Canadian mosquito goes by the scientific name of *Aedes vexans*.

A program of wholesale slaughter of mosquitoes was willingly supported by the taxpayers of Yellowknife between 1984 and 1988. It was designed to wipe out most of the mosquitoes within an eight-kilometre radius of City Hall. Each spring, all ponds, water-filled ditches and other suspected breeding sites within this area were mercilessly attacked. The sprayers came by helicopter, by foot and by dinghy, dumping over 20,000 kilograms of pesticides into the mosquito nursery beds. The idea was to kill the suckers before they could hatch.

Deserving at least an "A" for effort, the program was a net dud. Aerial spraying more often than not was foiled by gusty spring breezes. Ground equipment jammed. Field sampling showed that soon after the attack most lakeshores and ponds were still wriggling with larvae. Any adult mosquitoes caught in the cross fire were quickly replaced by other adults blown in by the whimsy of the winds.

The mosquitoes came. The taxpayers complained. City councillors decided to sell off its spraying arsenal and remaining pesticides. Perhaps they had realized that, in this wild and well-watered land, spraying insecticide to kill bugs is like trying to subdue a volcano with a garden hose.

Then came the summer of 1991. I made a point of asking old-timers if they had ever seen anything like it. "Never this bad," was the common reply. An unusually wet spring and high water levels contributed to a record hatch of mosquitoes. They became so thick that a friend living in a well-ventilated log cabin was obliged to erect a bug-free tent inside it to get some sleep. During one weekend in late June, over a dozen people were treated at the hospital for severe itching, inflammation or allergic reactions to multiple bug bites. Local retailers reported record sales of bug jackets, bug hats and bug dope. Dashing into one store on my way to a weekend adventure in the bush, I remember feeling a mild twinge of panic when I discovered an empty shelf where the bug dope used to be. "Sorry," said the cashier, "we should get another shipment in next week." Another twinge. *Next week?!* I prayed for unceasing wind in the meantime, the best natural shield against an all-out attack.

While at their peak, the mosquitoes made front-page news. We read about the possibility of an epidemic of insect-borne diseases and about the real-life tale of a woman driven half crazy by bugs while lost in the woods just 10 kilometres from downtown Yellowknife. Inevitably, the controversial spraying program crept back into the headlines. The apparent injustice at having to live with so many bugs prompted 2,000 residents to sign a petition calling for the reinstatement of the spraying program. The issue was put to a public referendum and the same 2,000 people showed up to vote yes. They won by a landslide. These events inspired me to submit a poem to the local newspaper:

> There once were mosquitoes in Yellowknife
> until they were voted away.
> Now it's time that we voted on winter
> it's long — can't we shorten its stay?

Heeding my scepticism and the exorbitant price tag for a new program — one estimate came to 50 cents per dead bug — City Hall did a quick turnabout and scrapped the whole idea.

Even if we could afford mosquito spraying programs — that *worked* — we would still have the blackflies, members of the Simuliidae family. The peskiest blackfly species breed only in flowing water, making their larvae largely immune to human meddling.

A constant turnover of cool, clear water provides the developing larvae with a continuous supply of oxygen and food. Lodged onto a rock or submerged plant, the wormlike larvae strain drifting plankton from the water. After a few days, they spin a pupa around themselves and metamorphose into the humpbacked flier we know so well. In early summer, pupa colonies are plainly visible, if you're inclined to look. But watch out. The newly formed flies may suddenly pop to the surface in a bubble of air and food will be the first thing on their little minds.

The taiga shield, with its countless rushing streams and rivers, offers ideal conditions for raising blackflies. Not surprisingly, they reach their greatest numbers here. Some people believe that this superabundance of blackflies may be a boon to preserving the region's wilderness character by forever limiting the number of people that make their home here. The renowned naturalist Ernest Thompson Seton, who explored this region in the early 1900s — and didn't stay — had little room in his heart for blackflies:

> The black flies attack us like some awful pestilence walking in darkness, crawling in and forcing themselves under our clothing, stinging and poisoning as they go.[95]

The journals of early visitors to this region are full of such accounts. Here's another from a 1930s prospector working out of Yellowknife. After listening to the "driving rain" of flies on his tent all night, he finally braved a dash outside to make breakfast.

In a few moments, the bacon in the frying pan was mottled with hundreds of black dots, flies that had perished. In the coffeepot floated a layer that covered the whole surface.[96]

As if disturbing your sleep, committing suicide in your food and flying persistently about your face or ears weren't enough, the blackfly is interested in stealing your blood. While the mosquito siphons up blood from below the skin, the blackfly gnaws out a surface crater with its scissorlike mouth, then laps up the blood that wells into the hole. Prime time for biting is when the weather is hot and humid and the air still. Like the mosquito, only the female bites, again to provide a protein boost to her developing eggs. She may take up to eight minutes to feed, often performing her work without you knowing it. She leaves behind a symmetrical round blotch that may last as long as a month, the universal tattoo of travellers in blackfly country.

On my first big canoe trip north of 60°, I made a special point of wearing only light-coloured clothing, which is supposedly repulsive to blackflies. Throughout the first day, I kept inspecting my apparel for roving flies. Not one. I figured that I had outwitted those so-and-so's. The trouble was that I didn't tuck my pant legs into my socks in true northern style. When I started rolling up my pants before wading over a soggy portage, I was shocked to see my lower legs covered in blackfly bites. The villains had sought out the darker corners of my flesh for a non-stop feast.

Fortunately for us, not all northern blackflies dine on humans. Many species are highly host-specific, feeding only on big game, small mammals or birds. One species, for example, goes only for loons, attracted to a specific chemical in its preening gland. They more than annoy birds and animals. They sometimes kill them. Cattle in northern Alberta have died within 15 minutes of a blackfly attack — from some sort of mysterious shock effect, not the loss of blood. Blackflies are more likely to cause death or serious discomfort in wildlife through their role as a vector for parasites. For instance, many a duckling has died due to a malarialike parasite transmitted by a blackfly as it bites.

A range of biting flies can sometimes make life miserable for large mammals, particularly caribou. In the summer they may be so plagued by blackflies and bulldog flies (Tabanidae family) that they dash across the tundra with such recklessness that some animals may be seriously injured, even killed, by a nasty fall or a collision with a rock outcrop.

The warble fly, *Oedemagena tarandi*, is especially annoying to caribou. While buzzing about their legs, it deposits eggs just above the skin. These hatch into sawtoothed larvae, which burrow into the leg and make a long journey to the animal's back. Here they ride all winter in fibrous sacs just under the hide. In spring, they climb out through their breathing holes, fall to the ground to pupate, then emerge as flies born with a clear search image for caribou legs. And so the cycle goes on.

Nose botflies, *Cephenemyia nasalis*, harass caribou by flying incessantly around their nostrils while depositing larvae. After a brief but irritating stay in this part of the animal, the larvae migrate to their winter home in its throat. In high concentrations (150 plus), botfly larvae can make breathing difficult for a caribou, especially when it has to run for its life with a wolf at its heels.

The relationship between these insect pests and their host is so intimate that caribou have evolved certain protective movements — besides running like crazy — that minimize infestation. They often stamp their feet, twitch their hides and shove their heads close to the ground with their nostrils buried in a low bush or clump of sedges. When the bugs get really bad, hundreds of caribou may band together in a closely packed mass, with only the outer animals exposed to fly attack. The unlucky ones at the edge constantly squirm their way into the protected centre, leaving others to take their turn in the war zone.

If only we could somehow do away with mosquitoes and biting flies altogether, for the well-being of humans and animals alike. Forget it. There is a bright side, however. Being the most abundant form of wildlife north of 60° (their combined biomass would exceed that of all the caribou herds put together), these insects play a vital role pumping energy into northern food chains. As aquatic larvae, they provide food for many species of fish and ducks. Insect-eating birds such as swallows, flycatchers and warblers absolutely depend on them. They also help to satisfy the voracious appetites of dragonflies and damselflies, which in turn become food for all kinds of birds and the occasional frog. And in the process of foraging for nectar, they serve as the main pollinators for innumerable northern plants, including some berry bushes, a favoured source of food for many animals, both four- and two-legged. Keep these benefits in mind the next time you are besieged by bugs. A positive attitude towards your attackers may serve as a last line of mental self-defence when all else fails.

Much more benign than mosquitoes and biting flies are the springtails. In fact, they usually cause no harm to any living thing. Most are saprophages, eating only dead material that falls to the ground.

More kindred to lobsters than blackflies, springtails are not true insects. They are members of another arthropod order, the Collembola, known mostly by their multisegmented bodies and pogo-stick tails. Latched like a mousetrap under the springtail's belly is a long forked structure that, when released, can send it sailing over distances hundreds of times its own pencil-point length. If humans could jump to the same proportions, topping a 12-storey building in a single bound would be no problem.

We owe a lot to springtails. Without them, the soil around here would be even sparser than it is now. The density of springtails hard at work in the soil is astounding, ranging from hundreds to tens of thousands per square metre. They are key players in the process of recycling nutrients and making soil. Their role in life is to grind up dead vegetation and other organic matter into minute particles that *only then* can be attacked by fungi and bacteria, which carry this process to completion.

If you've never seen them, snow fleas could easily be lumped in with unicorns, centaurs and other creatures that dwell only in the imagination. But they do exist, the name being ascribed to a type of springtail that comes out from under the snow in droves around late March or early April. This is mating time for snow fleas.

I once watched hundreds of snow fleas appear in my ski tracks as if out of thin air. While pondering their seemingly aimless hopping, I thought about their unusual sex life. Whatever wooing and strutting may go on down there on the snow, it never culminates in embrace. The males and females may not even see each other, yet fertilization takes place. At some magic moment, the male secretes a packet of sperm on a tiny stalk shaped like a golf tee. When the female comes along, she scoops up the sperm packet and inserts it to fertilize her eggs. A few weeks after they are laid, the eggs hatch on the warm, snow-free soil. The newly emerged nymphs get down immediately to the business of soil building.

Any sign of new life this time of year is exciting. Looking for mating swarms of snow fleas ranks high on my list of spring rituals. Discovering them assures me that winter is well on its way to oblivion. For this reason, I believe that the emergence of snow fleas merits wider recognition. Up here, as the snow begins to melt, we have no groundhogs out looking for their shadows. In their absence, why not celebrate "Snow Flea Day"?

Combine all the insects, spiders and springtails found in the taiga forest and you end up with around 10,000 species. That number drops off quickly as you move northward out of the trees. Much of their world remains a mystery. About a dozen species of biting insects have stolen most of the attention of entomologists venturing north. The complete life cycle of only a few species is known. How they cope with long periods of subzero temperatures is a subject of continued debate. New species are still being discovered. To address this dearth of knowledge, much research, including some very basic natural history work, must be done.

Don't wait for scientists to unveil the secret lives of northern insects and their kin. There are too many species and too many secrets. Equipped with a camera, a hand lens or simply a curious eye, go do your own studies. Gently probe through the surface of the forest floor. See what's moving behind the bark of a burned-out tree or under a cushion of moss. Park yourself beside a marching column of ants and see where they've come from and where they're going. Be prepared for some fascinating discoveries. But don't forget your bug dope. You never know who might be coming in for a landing.

14

Fire:
The Great Destroyer and Sustainer

May 24: From 6,000 metres up, the pilot of a 727 jet en route from Norman Wells to Yellowknife spots the distinctive grey plume of a forest fire just east of Rae Lakes. He reports the fire to Yellowknife's control tower, which relays its approximate coordinates to the Regional Fire Centre in Fort Smith. A reconnaissance plane is promptly dispatched to pinpoint the fire's location and assess the need for fire attack. Smoke haze from several large fires in northern British Columbia and Saskatchewan obscures visibility and the Rae Lakes fire cannot be found.

Staff at the fire centre are surprised but not alarmed at the number of forest fires already burning. According to conventional wisdom, early-season fires so soon after snowmelt do not get out of hand. Meanwhile, the winter's moisture is rapidly sucked out of the forests by several days of 25°C temperatures. Strong, dry winds from the south fan the flames of hidden fires.

May 28: The Rae Lakes fire is reported by another pilot, but again its position cannot be fixed due to thick haze.

June 2: Not one, but two forest fires are located east of Rae Lakes, both of them started by runaway campfires. The largest one is galloping northward between Margaret and Seguin lakes. The fire covers about 1,800 hectares when 36 fire fighters arrive on the scene. Whipped by gusts and whirlwinds from nearby thunderstorms, the fire nearly doubles in size over the next three weeks, before they can bring it under control.

June 20-22: An intense low-pressure zone over the Western Arctic lets loose thousands of lightning bolts, igniting 110 forest fires over a period of only 48 hours. Staff at the Fire Centre scramble to assess the damage and sort out fire-fighting priorities. Efforts to locate new fires, let alone fight the known ones, are hampered by a ubiquitous pall of smoke.

Separate fires burn into each other, their flaming halos merging, amoebalike, into one. Fifty kilometres north of Yellowknife, six fires centred around Rocky Lake run together, creating one of the region's largest, dirtiest conflagrations on record.

Over the next two months, the Rocky Lake fire burns 85,000 hectares of taiga forest — about the same size as metropolitan Toronto. Over 80 fire fighters are sent to the fire line, collectively putting in the equivalent of two years' worth of digging and chopping, flying and spraying. When the operation is over, the bill for this one fire comes to $100,000. As the smoke clears, patrol planes circle the perimeter of the burn looking for potential flare-ups. From the air, pilots look down on a butterfly-shaped patch of charred landscape that stretches 50 kilometres along both sides of the Yellowknife River.

July 20-22: Lightning from another series of thunderstorms ignites 40 new fires, several of them threatening to burn down the wooden power poles along the Snare River hydro line.

August 15-20: Yet another storm system moves straight north out of Alberta, this one dousing the Great Slave area with over 100 millimetres of rain. Winds gusting above 90 kilometres an hour knock out the Fire Centre's main microwave tower, bringing a healing three-day silence to their overworked communications team. Thanks to all the rain, fire safety signs along the Mackenzie Highway, for the first time this summer, warn of "Moderate" rather than "Extreme" hazard.

The respite from fire fighting is short lived. Little rain falls on the taiga forest east of the Slave River, the so-called Caribou Range. It is still bone dry and ripe to burn. As the storm system over Great Slave Lake subsides, a new one develops, heading east towards the tundra and throwing down more sparks than rain along its path. In the wake of this storm, some 20 more fires spring up in the Caribou Range alone. They advance as far east as Ennadai Lake, burning a total of over 12,000 hectares before the rains of late August finally snuff them out.[97]

* * * * * * * * * *

The year was 1973, a record buster for area burned, fires fought and dollars spent. No two fire seasons are alike. Each develops its own individual signature determined by the dryness and species composition of the forests, the temperature and humidity of the air, the strength and direction of the winds, the severity and extent of thunderstorms and the number of humans sitting around campfires or chain smoking in the bush. Some seasons are a breeze, with frequent rains dousing most fires before they gather too much steam. Other seasons, like 1973 and, even worse, 1989, transform the western Northwest Territories into a war zone.

To compare fire fighting to war fighting is only natural. Both depend on the capability to thwart the opponent's chances of surprise attack, to keep a constant surveillance on its activities and movements, and to send, at any moment, highly trained forces plus massive amounts of equipment and supplies to the front lines of battle. Words like "onslaught," "attack" and "retreat" are commonplace in the annual fire reports written after a burn is declared officially "out" by the brass at the Fort Smith Fire Centre. If, like soldiers, fire fighters were recognized by rank and awarded extra stripes for experience and valour, then people like Ray Schmidt would surely be colonels by now. For over 25 years he has battled with northern fires — on the ground, from the air or, more recently, in front of a computer screen. "I got tired of chopping spruce trees with a helicopter. Now I'm pretty much strictly in the fire detection business."

Before retiring in late 1991, Ray told me how things in this business had changed since the early days of his forest fire career. Back in the '60s and '70s most detection was done by aircraft. Luck and the goodwill of observant pilots played a large role in this game. Then in 1978, the Fire Centre added electronic eyes to their surveillance equipment by installing a lightning position analyzer (LPA), a device used by weather forecasters to track the progress and direction of thunderstorms.

"When they turned on the switch, the results were astonishing! We never realized that there was that much lightning out there." Some summer mornings, Ray would come into work to discover an LPA record of more than a thousand lightning hits overnight. "It gets frightening. But fortunately they're not all going to start fires. A lot get rained out before the flames really get going."

Lightning has been the main cause of northern forest fires since trees first established themselves after the last glaciation. Since then, every square kilometre of forested land on the taiga shield has probably been burned over at least 40 times. Evidence of fire's destructive impact on today's forests is everywhere. Fire-scarred trees and tottering snags are present in forests of all kinds and all ages. Charred stumps and branches litter the ground. And below the forest floor, charcoal or ash occur universally in the soil.

The most destructive burns develop when the tremendous heat from a large fire begins to exert control over local weather patterns, particularly the winds. As a large bubble of warm air rises over the fire, cooler air rushes in from all sides to fill the vacuum. Fuelled by these strong indrafts, the fire's rate of burning may suddenly increase by 500 percent. Powerful convection currents carry smoke and ash high into the atmosphere, forming a huge black mushroom cloud, the tell-tale sign of a monster fire. Fire bombs carried by turbulent winds rain down on unburnt forests kilometres away, starting new spot fires that are soon enveloped by the mother fire's quickly spreading ring.

At the flaming front of such monster fires, stands of trees may be literally vapourized by intense radiation. In a matter of moments, solid wood is transformed into carbon dioxide, steam and ash. Thin organic soils are burned down to bedrock. Silt and fallen debris clog creeks and streams, smothering the developing eggs of walleye and whitefish. Newly hatched bald eagles are torched in their nest. In the fire's aftermath, grey ash settles in gullies, forming drifts a metre thick.

To witness an intense fire at close range — I mean so close that you can *really* feel the heat — is to experience a holocaust, an apocalypse or maybe hell on earth. In terms of energy released, forest fires are comparable to other catastrophic events in nature such as volcanoes, earthquakes and typhoons.

Worst-case scenarios aside, the image of fire the destroyer is well deserved. The creators of Smokey the Bear have taken this perspective to its outer limits, providing us with vivid images of panic-stricken birds and animals running headlong into the fire and moonlike landscapes scorched beyond recovery.

But the land invariably does recover. The scars heal. Life returns. Ironically, it is fire that *sustains* the taiga forest's long-term health and vitality. Without fire, it would become unproductive, stagnant, ecologically monotonous. Fire prunes out aging forests and rejuvenates the landscape. Like the mythical phoenix bird that rises anew from its own funeral pyre, a new forest rises from the ashes of destruction, its soil enriched, its vigour restored and its floral and faunal diversity greatly enhanced.

Fire provides a shortcut to the release of stored nutrients into the soil, an especially valuable role in northern forests, where the decay of organic matter is ploddingly slow. Fire opens up the canopy of a forest, which lets more sunlight, heat and moisture fall on the forest floor, promoting the growth of saplings and shrubs. Fire further promotes regeneration by helping to clear away tangled branches, deadfall and needle litter that can choke out new plant growth. Fire, if hot enough, can cause a lowering of the permafrost table, providing tree roots with more room to seek out nutrients and support. Fire purges old growth forests of insect pests and disease. And fire boosts the overall productivity and variety of habitats available to birds and mammals.

Fire shows its handiwork most clearly in the pattern of forest disturbance left in its wake. The cliché image of homogeneous spruce forests stretching from horizon to horizon does not hold for the taiga shield, or for any region of the boreal forest where wildfires still burn. Foresters on the shield of northern Manitoba discovered that the biggest patch of unbroken forest was about 4 hectares — that's it. Viewed from the air, the boreal forest presents a patchwork quilt of forest types that vary widely in species composition and age. This haphazard pattern often has little to do with differences in soils, topography or moisture regime. Rather, it is fire that gives the boreal forest this distinctive mosaic appearance, criss-crossing the landscape from different directions, at different times, in different intensities.

A series of fires roared over the land behind my River Lake cabin in the 1940s and 1950s, then again in 1971 and 1982. There are places where I can stand along the edge of a former burn with one foot in a young spruce forest and the other in an 80-year-old stand of trembling aspen. Elsewhere, a dense willow shrubland dotted with ancient jack pines flanks an open forest of mature birch; I can perform the same trick there, the boundaries are so sharp. A favourite watering hole for migrating ducks is called "Century Pond" by locals in honour of an isolated stand of 130-year-old spruce that borders one side of the pond. From the top of a bald hilltop nearby, I can see such relict islands of tall spruce in any direction, veterans of intense fires that felled most of the other big spruce in this neck of the woods.

What springs up after a fire depends a lot on the type of fire that has passed through. Ground fires can cover large expanses of forest without doing any apparent harm to existing plant cover. Smoke rising 10 metres away from an abandoned campfire — an all too familiar sight — is good evidence of a ground fire in the making. These fires are fuelled by peat, buried punky wood, tree roots and other organic matter below the forest floor. They may smoulder underground all summer — even all winter — and never break through to the surface. Or they may climb up a root and burst into flames, now fuelled by the upper layer of leaf litter, grasses, shrubs, tree saplings and deadfall. These surface fires may also creep up the side of a tree, occasionally torching an entire tree, or worse (depending on how you look at it), setting off a crown fire, the wildest of wildfires.

Once a fire gains access to a tree's crown, the main canopy of branches, it can spread easily to other trees, often moving faster than a person can run. A crown fire may burn intermittently, with its rate of spread controlled by the advancing surface fire below. Fanned by a stiff breeze, it may race ahead of the surface fire as flames leap sideways from one treetop to another, leaving the ground below unscorched. In the classic crown fire, flames from the surface fire and upper canopy merge, forming a well-defined wall of fire that might extend 30 metres into the air. Churning balls of fire can suddenly take shape above the fire wall as hot cells of volatile gases ignite in the higher, more oxygen-rich air. In the path of such a crown fire, all but the wettest plant communities go up in flames.

When the smoke finally clears after an intense fire, the landscape may appear as desolate as the moon. I have seen a vigorous black spruce–feathermoss forest transformed into a hummocky grey-black wasteland punctuated only by bristly snags and heat-shattered bedrock. Large patches of rusty-coloured mineral soil lay freshly exposed in spots where the fire had burned off the overlying moss and peat. Flying over the site two years later, I had to do a double take: the entire landscape was covered by an unbroken pink carpet of fireweed. On the ground I discovered that those apparently lifeless clumps of charred cones topping some of the snags had in fact contained viable seeds. Tiny black spruce seedlings were popping up everywhere.

Plants that produce large quantities of seeds and spread easily in the wind have a free-for-all on freshly burned sites. Chances of seed germination and seedling survival are best where organic layers have been burned off and mineral soil has been exposed. This gives plant pioneers such as fireweed, willows and poplars a head start at colonization, since they take root most quickly in mineral soils laid bare by severe burns.

The more intense the fire, the more likely black spruce will become the dominant tree species, since it too regenerates best in mineral soil and its cones can survive much hotter temperatures than its chief competitor, the white spruce. To a large extent, the black spruce *depends on* fire for reproduction, since its serotinous cones release seeds most readily after they have been licked by flames. On favourable ground open to full sunlight, young black spruce can grow at a prodigious rate — up to 30 centimetres a year. By producing seeds much sooner than most other trees, black spruce can spread rapidly, maintain their dominance over other species and provide guaranteed life insurance for the next generation. Collectively, these traits make black spruce one of the most successful post-fire colonizers in the taiga forest.

Jack pine is another "fire species." In this region, no tree catches and carries a flame as well. Its loose, scaly bark, highly resinous needles and outstretched branches invite the spread of fire. Like the black spruce, many of its cones are opened by fire. The jack pine dominates the most exposed habitats — dry, rocky ridgetops and sandy upland plains — where fire tends to spread most quickly and visit most often. Once established on its preferred habitat, the jack pine is unlikely to give up its turf to other tree species so long as regular fires keep cycling through the forest.

An intense fire is the jack pine forest's key to self-perpetuation. For poplar and birch forests, it often spells doom. Fortunately, fuel buildup in deciduous stands is usually too light to carry a hot, persistent fire. Like the parting of a tidal wave, a full-scale crown fire may detour around such stands. Or it may become locally subdued, merely singeing the surface as it pauses on its course towards more combustible coniferous fuels. Even if every last tree is killed by a surface fire, a new deciduous forest will sprout from the ashes if root suckers of poplars and dormant buds around the trunks of birch trees have been spared from intense heat.

In the wake of a forest fire, there are no clear winners or losers among plants. The intensity of the fire, the type and extent of forests burned, the availability of seed sources for regeneration and the local topography and moisture regime all interact in complex, often unpredictable ways to determine the sequence of plant communities that develops after a fire. It's the same for wildlife populations. Each fire influences their distribution and abundance in its own unique way.

No doubt a few animals and birds are devoured by flames, as Smokey and his gang would have it. But the effects of direct destruction of wildlife by fire are insignificant compared to the changes it brings to their habitat. Spruce grouse, red squirrels, marten and other species that thrive in old growth spruce forests may show a marked population decline immediately after a large, hot forest fire. On the other hand, unless everything is uniformly scorched, some animal populations remain basically unchanged by fire. For instance, field studies of mice and voles during and after a controlled burn have demonstrated that they can usually avoid the worst of a fire's heat by burrowing, hiding in rock outcrops or seeking out patches of unburned forest. Once the fire passes, their populations may carry on as usual, showing little fluctuation up or down.

By far, the typical response of most kinds of wildlife to fire is to increase. Fire opens up old, stagnant forests and triggers vigorous new plant growth: tender young trees, a wide variety of shrubs and succulent grasses and herbs. With the bloom of youth restored to the landscape, herbivores are the first to benefit. Beavers depend on trembling aspen, birch and other deciduous species that spring up after a fire. Moose are notoriously fond of willows. So are snowshoe hare, ptarmigan and ruffed grouse. Renewed growth of herbaceous plants benefits red-backed voles, muskrats and seed-eating birds such as sparrows and redpolls. In turn, as the plant eaters prosper, so do their predators — wolves, foxes, lynx, weasels, mink and a long list of raptorial birds. And who has not witnessed the explosion of berry bushes after a fire without thinking of bears? They too benefit from the enriching touch of fire on the land.

Ah, but what about barren-ground caribou? Is it true that fire on their winter range, the taiga forest, is among their worst enemies? Have caribou starved to death in some parts of their range due to the widespread destruction of lichens, one of their primary foods? In the early 1960s, a series of intense fires swept across the western taiga shield, burn-ing over two million hectares of forest traditionally used as winter range by the Bathurst and Beverly caribou herds.[98] What impact did this have on these animals? Did it cause their numbers to decline? Was their migration subsequently deflected to other areas?

As far as I can tell, no one has a clear answer to any of these questions. The fact is that while forest fires continue to sweep across their winter range, caribou remain as healthy, as abundant and as unpredictable as ever.

Lightning strikes a tinder-dry 90-year-old white spruce tree, igniting loose scales of bark around its trunk. Branches just above the growing flames boil off the last of their moisture and release an invisible cloud of explosive gases. Just as the expanding fuel cloud envelops the tree's uppermost branches, it is licked from below by a flame. The entire tree ignites in seconds. Half an hour after the strike, a hectare of forest is in flames. Within two hours, the fire has grown 30 times bigger. By the time heavy rains douse the fire six weeks later, it has charred 80,000 hectares of shield country.

Two springs later, the sound of flickers drilling for insects reverberates through a forest of lifeless snags. Jack pine, birch and willow seedlings dot the bare, sandy ground. After five years, many low-lying areas once dominated by mossy hummocks and stagnant water are now deep beaver ponds rimmed with trembling aspen and alive with muskrats, mink and nesting ducks. Twenty-five years after the fire, willows are thinning out as they give way to increased shade from the closing forest canopy. A thick carpet of lichens and moss begins to dominate the forest floor. Sixty years after the fire, black spruce are well established below the pine and birch trees. But before they can overtake their taller hosts, a crown fire hits, beginning once again the eternal cycle of forest destruction and regeneration.

So goes one of an infinite number of possible scenarios triggered by fire. Though on average only 1 percent of the Northwest Territories' forests burn each year, the effects of fire are pervasive and long lasting. Fire keeps the region's vegetation cover in a state of constant flux and readjustment. It is the key ecological mechanism that initiates the growth of new forests, determines their species composition and age structure, and defines habitat options for wildlife. Taiga species have evolved in response to wholesale disturbances wrought by fire. Plants have developed strategies for rapid growth and reproduction that keep their life cycles in tune with the average frequency of forest fires. Many birds and mammals depend on the mosaic of vegetation communities created by fire to meet all their needs for food and shelter throughout the year.

In short, the taiga forest is a fire-dependent ecosystem. It's been that way for millenia. And as long as lightning flashes over northern skies and spruce pollen blows in the wind, it always will be. As a Dene trapper once said, "Whoever created the world, created lightning. If lightning starts fires, he's doing it for a reason."[99]

Part 2:
A Tapestry of Habitats

If you don't know who the Creator is, just go outside.
— Joe Boucher, *Dene Kede*[100]

15

Rock Outcrops: Bald and Beautiful

Good-bye industrial heartland of Canada. See you next family reunion.

Often when I travel to different parts of the country I like to bring home mementos of my visit. Not pennants for my kids or stickers for my car but *objets d'art* provided by mother nature. Once, on my way back from Whitehorse, I had a slight altercation with an airline attendant who had some misgivings about one such object. It was a metre-long piece of driftwood from the Carcross sand dunes. It had a pleasing shape and interesting cracks and crevices packed with sand. The attendant let me through after I told her that it was a "botanical specimen." This trip back it's a chunk of limestone pilfered from the north shore of Lake Erie. I take one last fond look at it before stuffing it into my suitcase.

"You're taking rocks *up there*?" asks my mother-in-law. This is a reasonable question given that around Yellowknife naked bedrock outstrips anything you could call forest by about 10 to 1. It is a geologist's dream. "There's none like this," is my reply. The mottled gray and white surface of the limestone is studded with fossils — clamlike gastropods, corals and sponges. These creatures lived in the warm, shallow sea that covered most of North America 400 million years ago. Most of the limestone that once covered the northern shield was peeled away by glaciers. My intention is to place this specimen alongside a laughing Buddha at the bottom of my fish tank.

Our plane angles down towards the north shore of Great Slave Lake. From up here, the shield rock looks like blisters of grey paint bubbling up through a thin veneer of green. This pattern disintegrates along the water's edge into a constellation of islands. Along some stretches of the north shore, there are over 100 islands per square kilometre. Splinters off the shield. A silvery orange light reflected on the water shows a washboard surface of waves kicked up by strong winds from the southwest. There is a distinct "V" pattern behind each island, like the wake of a boat. The islands appear to be moving towards us. I squint my eyes. They become a flotilla of birch-bark canoes coming to greet the ship. Welcome back to shield country.

Twenty-four hours later, my ship is a kayak, my destination Pillow Island. By convention, this island has no name. On the Geological Survey of Canada map, all that's shown for this blip in Yellowknife Bay is that it is made out of sedimentary rock. Now, my respect for the GSC knows no bounds. If it weren't for this crew, the region's geology might still be a total mystery (rather than a partial one). Gold might never have been discovered here. Yellowknife might not exist. I might never have known this wonderful landscape. Nonetheless, this island, through and through, is not sedimentary rock but volcanic. And it's a great place to see volcanic pillows.

This island is clearly two-faced. Approaching it from the south side, I find no place where I might easily moor my kayak. Angular ledges and chunky boulders along the shore keep me paddling around the island in search of a more inviting place to land. From a previous visit, I know that the north side presents a completely different face, being relatively smooth and presenting few shoreline obstructions. I veer right.

The best view of pillows is along the island's eastern shore, where a few thousand or million or billion years ago, a clean vertical rock face was formed, now providing a glimpse into the island's inner anatomy. I stop paddling. A gentle evening breeze blows me towards the base of the cliff and nudges the kayak against the rock face. It is covered in dark round blotches, much as you would expect on the side of a giant toad, only smoother.

These are the pillows, formed when volcanic fissures beneath a Precambrian sea oozed molten rock. The lava flowed downslope in a tangled mass of saclike, bulbous lobes that fed the glowing edge of the volcanic pile. Advancing sporadically, they split into lateral buds or merged into each other to form wide-spreading tongues of lava. As new pillows piled up on top of older, more solid ones, their underbellies often took on a distinctive gravity-formed cusp, which today gives a good clue as to which way was up when they were formed.

I clutch the rock wall with both hands to keep my kayak from drifting and steady my view. Judging by the shape of these pillows, I conclude that what was once up is now almost down. Like so many of the rocks on the Canadian Shield, these have been tilted and generally knocked about into entirely new orientations since their creation. A lot can happen during half the lifetime of a planet.

I've seen pillows of all shapes and sizes (ever seen a "megapillow"?) in colours ranging from dark olive to golden brown to pink, and everything in between. They may be fine textured and smooth as pearls or coarse and fractured as old cement. Formed from different types of volcanic rock, primarily basalt and andesite, the colour and texture of pillows has been further influenced by how severely they were stretched, compressed or otherwise deformed over the eons. Somebody should do a field guide to volcanic pillows.

The pillows before me are rusty yellow with cinnamon rims. Some of them are dotted with jewel-like pockets of white quartz that flowed into tiny gas chambers within the cooling lava. Their outer rims have a rough, glassy texture, an artefact of rapid cooling as the lava was quenched by seawater.

I push off from the rock wall and head towards the north end of the island. Although I've plenty of sunlight left, I want to be sure to catch the right angle for photographing some other stories told by the rocks. After a few easy strokes, the north shore of Pillow Island comes into full view.

Sloping gradually out of the water are long, rounded humps of sculptured rock. In all of shield country, there's probably at least a million bedrock knobs like these. They are tapered in every way like the upper deck of a whale (minus flippers and fins), hence the name whaleback. Like its namesake, this feature is rarely found alone — geologists actually speak of "schools" of whalebacks. Sizes often vary within a school. They average about 4 metres across and 5 to 10 metres long. When they get much bigger, up to 100 metres long, they are called rock drumlins. The shorter ones might more properly be called "dolphin-backs." About a dozen of these rock mammals, big and small, swim up the northeast end of Pillow Island. The sculptor in all cases was glacial ice.

To say that a whaleback is smooth really depends on how you look at it. From a kayak on the lake, they appear smooth. A few centimetres from my nose and they are anything but. The rock is scratched and pitted and gouged like the surface of a freshly iced cake attacked by a fork-wielding child. Some of the scratches are less than a millimetre deep and only a few centimetres long. Others form deep channels in the rock, which I can trace for several metres.

These are glacial striations, engravings of the continental ice sheet as it made its southwards advance across the top two-thirds of North America. Anchored to its base were chunks of rock transported from "upstream" that were dragged, under unimaginable pressure, across the bedrock. One estimate of weight at the bottom of the ice sheet comes to around 3 billion tons per square kilometre — give or take a few million tons. Gazing at the abraded face of Pillow Island, I think of another day I spent out in the field with a geologist who was quite impressed by a similar display of glacial power: "What a great set of stris!" was all he kept saying as he walked back and forth over the lines. "What a great set of stris!"

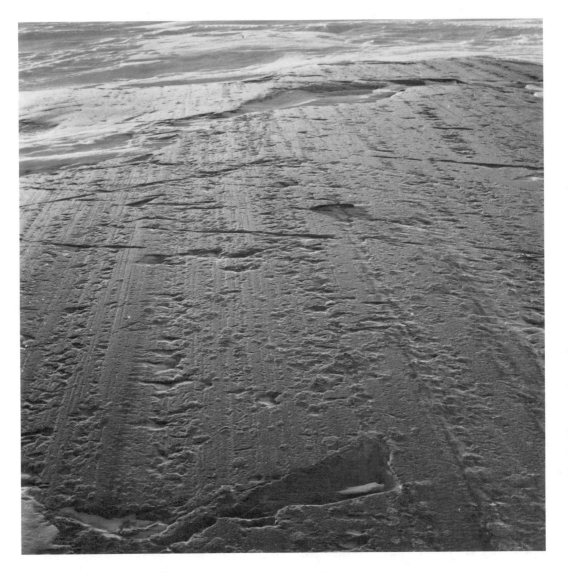

Glacial striations record the power and path of an advancing ice sheet kilometres thick. *(John Poirier)*

I examine the striations more closely and try to decipher their unique script. A few striations get off to a weak start, gradually become deeper and wider, then die out as if the engraving particle had been ground down, worn to dust or retreated up into the belly of the glacier. Other striations start abruptly, deep and wide, then fade away, as if a sharp side of the particle had suddenly flipped down onto the bedrock until it too wore down or retreated. Some show the reverse trend, starting thin and ending fat, suggesting that the pressure became too much for the engraving particle and something suddenly gave way — either the rock or the overlying ice. Most intriguing are the striations that come to an abrupt halt and are followed immediately by two smaller striations on either side. Here the engraving particle may have simply broken in two, with the little pieces picking up where the original one left off.

I hold a compass up to the striations. Their general trend tells me that, in this part of the world, the glacier advanced out of the east-northeast, or about 60° more or less. The trend of striations in different places provides a logbook of the ice sheet's journey across the land. For example, along parts of the Mackenzie Valley striations point to the northwest, while in the Liard Valley some angle to the southwest. These divergent trends suggest that the ice parted like a wave as it began to nudge up against the Mackenzie Mountains — some headed north and some headed south.

But what's this? Distinct sets of striations *in one level spot* that seem to head off in separate directions? It turns out that some of the deeper striations that appear to buck the general trend may be thousands of years older than their neighbours. In the last 2 million years, continental ice sheets advanced over the shield rocks at least four times. These aberrant striations represent the unerased record of glaciations that may have occurred thousands of years before the most recent ice sheet ploughed over shield country 30,000 years ago.

I follow some striations that diverge around a small knob on the side of a whaleback — perhaps a budding fin? Here the ice flowed like water as it passed this knob. The lines swerve from one side to another, then meet up again, much as a fast river makes way for protruding boulders. As unyielding and powerful as the ice may seem, this is evidence of its remarkable fluidity, often resulting when ice warms under great stress and flows like molasses.

With practice, reading striations for the flow of ice becomes like reading rapids for the flow of water. Other, less subtle signs on the rocks embellish this story. Mixed in with striations are irregular pockmarks and gouges made by good-sized rocks as they rolled along in a choppy ride underneath the ice. Chatter marks are an expression of such a ride. These are often displayed as a series of shallow, crescent-moon-shaped fractures created when a large boulder skips along the bedrock under fast-moving ice.

The sun gets lower, the shadows longer. The glacier's engravings become more pronounced, as if burning deeper into the bedrock. I clamber over the uppermost whalebacks and breathe in the view from the summit of Pillow Island. The sky is a tangerine blue, the water shining like mercury, the rocky shore a golden tan. A picture of mellowness. A linear U-shaped depression in the rock provides a perfect throne from which to watch the changing light.

Solar energy stored by the island's surface during the heat of the day now flows into my outstretched hands. The bedrock is silken beneath my fingertips. I am drawn to focus on it again. The depression snakes away from me on both sides and is uniformly smooth, with rounded edges. I stand up in excitement, realizing that I have just been sitting in a classic glacial groove.

This groove probably came into existence as a deep, rough-edged furrow in the bedrock surface ploughed by a large, tough chunk of rock at the bottom of the glacier. Once formed, this furrow could have acted as a convenient channel for fast-flowing currents of ice, which gradually smoothed and polished it into its present shape.

Glacial grooves usually form along points of weakness in the bedrock surface such as fractures or bedding planes where two types of rock meet. Here the groove seems to have sliced along the edge of two kinds of volcanic rock. In its dimensions, this is a fairly typical groove: about 15 centimetres deep, twice as wide and 15 metres long. In the softer sedimentary rock of the Mackenzie Valley, there are monster grooves, some over 25 metres deep and 10 kilometres long — so big that you need to fly over them in an airplane to comprehend their size.

Over thousands of years, glaciers have carved the bedrock of shield country in two basic ways. Whalebacks, striations, chatter marks and grooves all resulted from abrasion, the grinding down of rock by ice or objects embedded in the ice. Although extremely powerful, abrasion was a gradual process, removing small bits of rock much as a sculptor uses sandpaper or a file. This kind of erosion shows the glacier's more refined side, the results usually being relatively flowing or curving in appearance. Plucking was a much more immediate process in which angular chunks of rock, some the size of houses, were uprooted and bulldozed away by the advancing glaciers. While plucking, glaciers changed the shape of the land with all the finesse of a cosmic sculptor equipped with a sledge hammer.

Prime target areas for glacial plucking are fault lines, bedding planes between different rock types and clean fractures, called joints, through the bedrock. Water percolates into these zones of weakness and, through years of freezing and thawing, splits the bedrock into huge blocks ready for plucking. Meditate on one of the region's many crumbling cliffs littered with boulders. Or look down from a plane window at a typical series of clean-edged lakes that line up, nose to nose, along the same straight line. You'll understand the power of plucking, since this is how these signatures of shield country were formed.

A combination of these two kinds of erosion helped give Pillow Island its two faces. Abrasion left it smooth and sinuous on the upstream side. Plucking left it rough and blocky on the downstream side. I can see this effect in miniature on a sorry-looking whaleback that seems to have lost its head. Its tapered tail and body are intact — smooth, rounded and covered in striations like any whaleback. But this pattern ends abruptly in

A jumble of boulders created by the freezing and thawing action of water in the rock plus the plucking action of recent glaciations. *(John Poirier)*

a steep, jagged drop, and there are a few plucked boulders strung out in front. This island is a showcase! Here is a classic roche moutonée.

The smooth side gives this landform its name, *moutonée* being French for the curled wig worn by judges and lawyers in early European courts. Some geologists prefer the analogy to the back end of a sheep: *mouton*. Roches moutonées are found throughout shield country, some smaller than this one, some so big they are best appreciated a kilometre away. Most owe their existence to preexisting knobs that were probably sliced through with joints before the glacier came a-plucking.

Roches moutonées and other plucked landforms are one of the main calving grounds for glacial erratics. These are rocks that have been carried long distances by advancing ice. They may have rafted along on top of the glacier, been shoved forward by its nose or been carried below while entombed temporarily in its icy belly. Dropped a few metres or a few thousand kilometres from its source, an erratic often has little in common with the rocks around it — hence the name.

On the ragged side of Pillow Island I find small chunks of purplish brown shale, a type of sedimentary rock that probably travelled here from around the East Arm of Great Slave

Roche moutonée — a whaleback that lost its head. This feature clearly shows evidence of glacial abrasion on the upstream (left) side and plucking on the downstream (right) side. (*John Poirier*)

Lake. There are fragments of volcanic rock that probably weren't plucked from this island; their colour and texture are completely different. Parked right beside them are large boulders of granite. Some are pink with black flecks, others are white with pink flecks. Some are rough around the edges, as if freshly quarried from the mother rock, wherever that may be. Others, likely more well travelled, are polished smooth in places like half-finished tombstones.

I zero in on the pink granite with my hand lens. To my mind, this is a three-tonne gemstone. Suspended in a frozen broth of pink feldspar and black mica called biotite are glittering crystals of milky quartz, some the size of peas. As erratics or as bedrock, granite like this constitutes one of the most abundant rock types of shield country. Yet to see such crystals in the full light of day seems a privilege, considering that they originally formed as plutons many kilometres below the earth's surface.

Half-baked fudge and molten rock have a lot in common when it comes to crystal size. Cool the fudge quickly and it becomes a smooth taffy, its crystals fine, much like the air-cooled volcanic rock that spilled out onto the surface billions of years ago. Cool the fudge slowly and its crystals have a chance to grow, making the final product rough and

The ice sheet quarried this erratic from light-coloured granites and dumped it on an island dominated by volcanic greenstones. *(John Poirier)*

sugary, like maple sugar. In the same way, deep molten rock, insulated by a thick layer of solid rock above, became coarse grained, like this granite, as its mineral components had time to grow into discrete crystals plainly visible to the naked eye.

Granite is made up of three of the planet's most abundant elements: oxygen, silicon and aluminum. When oxygen and silicon bind together they form silicates, the hardest and most common of which is quartz. Stir in some aluminum and you get feldspar, which is almost as hard and as common. What gives these minerals their toughness is an atomic structure based on one of the strongest designs engineered by nature, a design copied by the American inventor Buckminster Fuller in his renowned geodesic dome: the tetrahedron. The basic building block of all silicates is a group of four equally spaced oxygen atoms locked into position around one silicon atom.

This simple but durable atomic structure explains many of the topographic ups and downs of shield country's present landscape. Rocks high in silicates are the slowest to erode and therefore tend to dominate today's skyline as knobs and rolling hills. These include not only granites but also some softer volcanic or sedimentary rocks that may have been shot through with molten quartz or other silicates after their original formation (for instance, Pilot's Monument, which towers over Yellowknife's Old Town). Silica content partly explains why, since the Precambrian era, the rocks of shield country have weathered and washed away at different rates. Glaciers gave this process a dramatic boost — probably 50 times or more during this period. The result is a landscape whose topography tells a fascinating story, whether viewed 1,000 metres up through a plane window or 10 centimetres away through a hand lens.

Like much of this landscape, the surface of Pillow Island was scoured clean by the last glacier. Today the island is still 80 percent bald rock. Ice continues to keep its shoreline free of soil or plant cover — not glacial, but lake ice grinding against it each spring during the first throes of breakup. Powerful summer waves also play a role in denuding the shore. Fire has probably burned off some of the thinner soils now and then. But one of the greatest forces keeping the rock outcrops of shield country naked is gravity.

Shield rock usually slopes in one direction or another, often at a steep pitch, and its smooth surface offers little resistance to the flow of water. It carries off particles of weathered rock, soil, needles, leaves and other organic matter — basically anything that's not either too big or too secure to budge. Small cracks, dips and hollows in the rock cause water to slow down and drop some of this material, providing a toehold for the growth of plants. During the driest part of summer, these areas also act as tiny reservoirs of stored moisture. As a result, vegetation on rock outcrops forms a tapestry of ribbons and pockets, reflecting, with stunning fidelity, subtle changes in the lay of the land.

The lichen pioneers reach their fullest expression on rock outcrops, forming a patchwork of crustose lichens on the steepest terrain and tuffet-like cushions on flatter surfaces. Soil often accumulates around these cushions, permitting the growth of various mosses, prickly saxifrage, pale corydalis, pussytoes and other herbs and flowers adapted to the dry, exposed conditions of rock outcrops. Several species of ferns found in no other shield country habitat also thrive in these conditions.

Scattered among the cracks, frost-shattered rock and glacial furrows are low shrubs such as ground juniper, creeping juniper, wild rose and kinnikinick. Isolated jack pine, spruce and birch can eke out an existence clinging to the tiniest of cracks, their roots extending sometimes several metres across bare rock until they dive into a moist depression. In the deeper hollows, thick mats of moss and lichens act as sponges for moisture, encouraging the growth of several plant species more typical of wetter habitats, including Labrador tea, mountain cranberry and cloudberry. Other berry producers such as gooseberry, blueberry and currant bushes often grow in these depressions, making them one of my favoured targets when fall harvest time comes.

Your chances of seeing a peregrine falcon, golden eagle or cliff swallow are greatest around steep rock outcrops, for this is their preferred nesting habitat. Ravens too nest here most often — that is, when they're not making themselves at home halfway up a microwave tower, on an office building roof or in the abandoned spruce-top nest of a bald eagle. In the form of exposed shoals and small islands, rock outcrops are prime nesting habitat for several kinds of terns and gulls.

As for other species of wildlife requiring more cover for reproduction and shelter, rock outcrops have little to offer. Perhaps the greatest universal value of rock outcrops to wildlife is their openness, providing a clear view of advancing predators and an unobstructed pathway for escape.

I too look frequently to rock outcrops for escape, not from predators but from the congestion of a young spruce forest, the sogginess of a bog or the insect chorus of the lowlands. My favourite hikes are along uninterrupted rocky shorelines, on top of long, exposed ridges or through chains of outcrops woven through the woods. Breaking suddenly out of the forest onto a bald patch of rock is one of the unique joys of travelling in shield country. The views are usually surprising, often stupendous.

While prospecting for valuable minerals along the bald eastern shore of Great Bear Lake in 1903, the geologist Mac Bell abandoned his quest for a moment upon discovering another kind of richness in the rocks:

> The high rocky walls were stained and weathered to beautiful shades of purple, red and brown, and gave, with the reflection of the precipitous cliffs in the clear northern waters, a singularly rich effect.[101]

Common Plant Species of Rock Outcrops[102]

Lichens
Iceland moss	*Cetraria islandica, C. nivalis* & others
Reindeer lichen	*Cladonia rangiferina, Cladina stellaris* & others
Rock tripe	*Umbilicaria* spp., *Lecidea* spp.

Mosses
Dryland mosses	*Polytrichum juniperinum, Hedwigia ciliata*

Ferns
Parsley fern	*Cryptogramma crispa*
Fragrant shield fern	*Dryopteris fragrans*
Polypody	*Polypodium vulgare*
Rusty woodsia	*Woodsia ilvensis*

Grasses
Bentgrass	*Agrostis scabra*
Reed-bentgrass	*Calamagrostis canadensis*
Blue Grass	*Poa glauca*

Trees
Jack pine	*Pinus banksiana*
Black spruce	*Picea mariana*
Paper birch	*Betula papyrifera*
Trembling aspen	*Populus tremuloides*

Low shrubs and ground cover
Ground juniper	*Juniperus communis*
Creeping juniper	*Juniperus horizontalis*
Crowberry	*Empetrum nigrum*
Kinnickinick	*Arctostaphylos uva-ursi*
Red currant	*Ribes glandulosum, R. triste*
Gooseberry	*Ribes oxyacanthoides*
Wild raspberry	*Rubus strigosus*
Wild rose	*Rosa acicularis*
Labrador tea	*Ledum decumbens, L. groenlandicum*
Mountain cranberry	*Vaccinium vitis-idaea*
Blueberry	*Vaccinium uliginosum*
Saskatoon berry	*Amelanchier alnifolia*

Herbs and flowers
Pale corydalis	*Corydalis sempervirens*
Prickly saxifrage	*Saxifraga tricuspidata*
Cinquefoil	*Potentilla nivea*
Fireweed	*Epilobium angustifolium*
Wild strawberry	*Fragaria virginiana*
Cloudberry	*Rubus chamaemorus*
Northern yarrow	*Achillea nigrescens*
Pussytoes	*Antennaria rosea, A. nitida*
Goldenrod	*Solidago decumbens, S. multiradiata*
Narrow-leaved hawkweed	*Hieracium scabriusculum*

Bird Species Commonly Nesting on Rock Outcrops[103]

Common raven	Herring gull*
Cliff swallow	Mew gull*
Golden eagle	Bonaparte's gull*
Peregrine falcon	Arctic tern*
(*near water only)	

16

Coniferous Forests: Forests of Little Sticks

Just as I cut my motor, a bald eagle lunges out from the top of a large white spruce near the water's edge and flies away. I drift into shore. The sound of aluminum on shale breaks the morning stillness. It must be 20°C by now, but my face and ears are still a bit raw from the chilly boat ride across McLeod Bay on the East Arm of Great Slave Lake. Though heat waves now shimmer off its dead-calm surface, there was rotten ice floating around out there just last week.

As I approach the spruce where the eagle had perched, I notice a tangle of sticks, bleached fish bones and downy grey feathers around its base. I look up to see the gargoyle-like head of an eaglet looking down at me from its treetop nest. A high-pitched cackle alerts me to its mother's return, and I give the nest tree a wide berth. I saunter along the boulder-strewn shore for a while until a gurgling creek lures me into the forest.

As if entering a cathedral, I step out of the sunlight into a hushed emerald world of towering white spruce. Like incense, the sweet smell of resin fills my nostrils. The forest is pleasantly cool and damp. I pad through a plush carpet of feathermoss, sinking to my ankles with every step. Some of the trees around me are over half a metre across and 15 metres high. I pull out my trusty Swedish tree corer and go to work on the biggest tree in sight — 195 years, give or take a decade or two. That puts its germination back into the late 1700s, about the time when Alexander Mackenzie was exploring Great Slave Lake en route to the Arctic Ocean. I try to imagine this forest in its youth as Mackenzie and his Chipewyan guides paddled by. Perhaps they stopped near this very spot for a shore lunch of lake trout and tea.

My reverie on the passage of time ends abruptly when a spruce grouse rockets out from under my descending shoe. It flies no farther than the nearest tree and lands on an exposed branch, making no effort to conceal itself. Such behaviour reminds me of the bird's wretched nickname, "fool hen," earned because it relies more on camouflage than flight to evade predators. I give it a nod and move deeper into the forest.

My main motive for today's outing is not gigantic spruce or birds — though always a treat to discover — but orchids. It's late June, the time of year when boreal orchids are in their greatest glory. There are about eight species on the taiga shield, and for many of them moist coniferous forests are the preferred habitat.

I wander, head down, through the forest, always keeping the creek within earshot so I don't get lost. Like fishing, an element of luck always helps in the quest for orchids. So it is today. I have this condition in which the call of an interesting bird can suddenly sidetrack me from the best laid plans. At the distant sound of two merlins chattering to each other, thoughts of both orchids and the creek slip completely from my mind. I am soon dashing over the feathermoss while trying to pull binoculars out of my pack. I arrive at a spot where a wind-fallen spruce has created a large light-filled clearing in the forest.

The merlins' nest must be nearby, since their chattering builds to a crescendo — though I never do see them. What I *do* see, basking at the edge of the clearing, is a magnificent patch of Richardson's lady's-slipper orchids. There are about a dozen of them, spread out evenly in a circular patch that might well have been planted by a Dutch gardener.

Looking through the wrong end of my binoculars, I zoom in on one particularly stunning specimen. (Try this technique sometime. It's very effective when you've left your hand lens at home.) Its flower droops coyly from a slender stalk, sporting a yellowish-green cap and an ovoid lip with a pale magenta tinge. A closer look reveals a row of royal purple dots inside the flower. My nose excites to its fragrant perfume. *Cypripedium passerinum* is what the taxonomists call it, the species name being derived from its lip, which has the shape and size of a sparrow's egg.

Today I'm twice lucky. I come to a sharp edge of the mature spruce forest — a boundary no doubt left by fire — and enter a more open forest dominated by much younger spruce and tall willows. My hunch is that, by moving into another kind of habitat, I'll find another kind of orchid. A purple flash on the forest floor proves me correct — a solitary bog orchis, *Orchis rotundifolia*. Surrounding the central flower are two narrow petals that point out like the ears of an alert dog. The pronounced lower lip is marked with parallel rows of purple splotches, which act as landing lights for bumblebees and other pollinating insects.

I take a few pictures. Then, prompted by my latest success, I continue looking for other species, with an image of the celebrated Calypso orchid foremost in my mind. Instead of more orchids, I discover wintergreens, another group of showy flowers found most often in coniferous forests. Showiest of them all is the large-flowered wintergreen *Pyrola grandiflora*. I find a patch of them around the base of a veteran spruce. Like the orchids, these flowers are at their radiant best this time of year and I take several minutes, lying prostrate in front of them, to admire their creamy white perfection.

The midday heat reminds me of my vow to take a plunge in the lake, no matter how cold it might be. It gradually dawns on me that I have no idea where the lake is, let alone the creek that was supposed to guide me back to my boat. A red squirrel decides to choose this moment to go into its scolding routine, hurling abuse at me from a nearby branch while gyrating its body and tail like a Chinese windup toy. A blackpoll warbler follows suit by letting loose a rapid series of lisping "tisk-tisks."

In just about any spruce forest, the chances of finding solitude and getting lost are equal. The solution: get high, preferably in a tree or on a rocky hill. Seeing uniform flatness in all directions, I opt for the former route and find an isolated jack pine that sticks up above the rest of the trees. While climbing, I think of this species' western cousin, the lodgepole pine, which goes by the scientific name of *Pinus contorta*, in recognition of its twisted and tangled growth habit. My jack pine is no less contorted and its jumbled branches make climbing a challenge, to say the least. I finally rise above the forest and am rewarded with a stupendous view of McLeod Bay, still calm and still shimmering.

At the peak of a neighbouring spruce, I spot a ruby-crowned kinglet making brief aerial forays into a swarm of hapless mosquitoes. It stops between courses to reestablish its territorial rights, sending forth a bellowing warble, which, to me, has always seemed disproportionate to its petite stature. But who am I to judge what song this bird should sing? The kinglet makes its home in the treetops. It belongs here and is complete. As for me, my badly scratched legs are giving out and my lunch is back down on the ground.

The shore is easy to find. My boat is not. With no boat in sight, I'm not sure whether to walk right or left. Numerous bays and subtle curves obscure my view. Perhaps I could swim just a bit from shore to get my bearings again. I cup my hands in the lake for a drink and a temperature check. Swimming is no longer an option. Think James. Flip a coin maybe.

Suddenly I hear the familiar cackle of a bald eagle. It approaches me, flying low, just over the trees. Having apparently sized me up, it turns slowly back towards its nest. As it pivots effortlessly in the air, I marvel at its dark, massive wings that span 2 metres from tip to tip. Unhesitatingly, I follow the eagle, feeling grateful that it showed me the way home.

When I first came north as a graduate student, my job was to explore different techniques for classifying the landscape into neat, manageable packages. With this in mind, I had the notion that it would be easy to go out into the field with my camera and photograph a typical spruce forest — I mean the *classic* spruce forest of the Canadian taiga. In preparation for my first field season, I spent much of the winter trying to impose order on a landscape I had never seen by drawing squiggly lines on photos taken from the air and infrared images beamed down from a satellite. This mind-numbing work told me that all spruce forests looked pretty much the same.

After a few weeks in the field and several rolls of film, I abandoned my quest. I discovered that, although spruce forests do appear similar from a distance, closer inspection reveals some profound differences. I also learned that the other coniferous species of this region, jack pine and tamarack, had a greater presence on the ground than my fancy satellite images had let on.

Picea mariana, that water-loving, acid-tolerating, quickly propagating black spruce, dominates most of the region's forests. Just *how* dominant it is can't be represented by a number. Does it make up 75 percent of the taiga forest? 95 percent? Nobody really knows. Put it this way: if you're in doubt, it's black spruce.

Black spruce forests growing in the soggiest, lowest areas are characterized by one other type of plant: sphagnum moss. In the black spruce–sphagnum forest, permafrost is usually widespread and close to the surface, thanks to the insulating effect of the moss carpet. With permafrost so extensive, the annual freezing and thawing of the active layer causes much commotion in the soil. As a result, hummocks are everywhere. Low shrubs of Labrador tea, dense pockets of mountain cranberry and occasional patches of *Cladonia* lichens usually crown the hummocks. Many are skirted by sedges, cotton grass or other plants that can handle moister conditions. Tamarack grow on the wettest sites, often where the forest gives way to true wetlands. But this species usually grows alone or in small, isolated stands, being no match for the adaptable black spruce.

Spruce forests on drier sites may look the same, but they are fundamentally different in several ways. Feathermoss, not sphagnum, carpets the floor. Low shrubs are much sparser, if not lacking completely. The underlying peat may be only a few centimetres deep. And the ground is much less hummocky. Sphagnum may be present in the spruce–feathermoss forest, but only on isolated hummocks. Horsetails, cranberries, bearberries, bunchberries and wintergreens are common plants growing on the feathermoss. Their coverage is relatively light where trees grow closely together, blocking out much of the sun's life-giving energy.

Either species of spruce may dominate this kind of forest, depending on how well water drains from the soil. White spruce find soggy soils intolerable. Unlike black spruce, their nutrient uptake in such conditions is seriously handicapped and their roots may drown for lack of air. But on well-drained sites, such as lake shores, river floodplains, old beach ridges or hillsides covered in glacial till, white spruce trees hold the clear advantage. Anywhere else this kind of forest occurs, black spruce retains dominion.

Next time you're over shield country in an airplane flying northeast towards the tree line, keep an eye out for a curious change in the forests below. It's not the tree line you're looking for. This change happens long before the trees start disappearing. Bedrock outcrops become scarcer and the landscape looks more streamlined, reflecting a thicker blanket of glacial deposits. Spruce become more widely spaced. Many are quite tall and, being open grown, have the symmetry of well-manicured Christmas trees. Most remarkable is the change in the forest floor. It gradually brightens and becomes dominated almost exclusively by lichens. Welcome to the spruce–lichen woodlands.

Sandwiched between stunted patches of tree-line forest and the more closed forests along the shield edge, the spruce–lichen woodlands form a large-angled swath of relatively even forest extending from one end of the region to the other along a southeast-northwest axis. They grow mostly on drier, gently domed plateaus where the veneer of glacial till is thick and peat is thin. Again, black spruce dominate the forest, though they still concede to white spruce on better drained ground. Jack pine and birch often mingle with the spruce, especially in younger forests. Sharing the forest floor with over a dozen species of lichens are scattered mats of Labrador tea, shrub birch, crowberry and two species that alone provide reason to visit — the showy alpine azalea and the ever-sweet blueberry.

Finding a wayward ball at the Yellowknife golf club is rarely a problem. The jack pines along the fairway (if you can call it that) have few lower branches and the undergrowth is sparse. Low shrubs such as soapberry, cranberry and kinnikinick, plus a few scattered patches of grass, lichens and bald sand, provide little cover in which a day-glow ball can hide. Mature jack pine forests have a parklike aspect and usually provide welcome elbow room and dry footing if the going gets too rough while travelling through more congested, soggier forests. They reach their finest expression on dry ridges and flat benches where there's lots of sand or gravel. The most exposed rock outcrops with the thinnest soils are another jack pine stronghold.

Fire almost guarantees the birth of a new pine forest, since the seeds of this species are released by flames and seedlings grow rapidly, usually outstripping their competitors. After a few years, countless saplings clog the ground in a battle for ascendancy. Unlike its mature version, the pine forest at this point can be hell to walk through. Certainly golf is out of the question.

Common Plant Species of Coniferous Forests[104]

Lichens
    Tree hair   *Alectoria sarmentosa, Usnea* spp.; *Bryoria* spp.
    Trumpet lichens   *Cladonia coccifera, C. chlorophaea, C. pyxidata* & others
    Shrubby lichens   *Cladonia uncialis, C.cornuta, C. amaurocraea* & others
    Leaf lichens   *Peltigera canina* var. *rufescens, P. apthosa, Nephroma arcticum*

Mosses
    Feathermoss   *Hylocium splendens, Pleurozium schreberi*
    Bog moss   *Sphagnum* spp.

Lower plants
    Horsetail   *Equisetum arvense, E. palustre, E. sylvaticum, E. scirpoides*
    Spikemoss   *Selaginella selaginoides*

Trees
    White spruce   *Picea glauca*
    Black spruce   *Picea mariana*
    Tamarack   *Larix laricina*
    Jack pine   *Pinus banksiana*
    Paper birch   *Betula papyrifera*

Tall to medium shrubs
    Green alder   *Alnus crispa*
    Willow   *Salix glauca, S. bebbiana,* & many others

Low shrubs and ground cover
    Ground juniper   *Juniperus communis*
    Crowberry   *Empetrum nigrum*
    Kinnickinick   *Arctostaphylos uva-ursi*
    Bearberry   *Arctostaphylos rubra*
    Mountain cranberry   *Vaccinium vitis-idaea*
    Wild rose   *Rosa acicularis*
    Shrubby cinquefoil   *Potentilla fruticosa*
    Soapberry   *Shepherdia canadensis*
    Labrador tea   *Ledum groenlandicum*

Herbs and flowers
    Twinflower   *Linnaea borealis*
    Lady's slipper orchid   *Cypripedium passerinum, C. guttatum*
    Bog orchis   *Orchis rotundifolia*
    Calypso orchid   *Calypso bulbosa*
    Northern bog orchid   *Habenaria obtusata*
    Northern comandra   *Geocaulon lividum*
    Bunchberry   *Cornus canadensis*
    Single delight   *Moneses uniflora*
    Large-flowered wintergreen   *Pyrola grandiflora*
    Short-stemmed raspberry   *Rubus acaulis*

Keep in mind that the forest types just described — the black spruce–sphagnum forest, spruce–feathermoss forest, spruce–lichen woodland and jack pine forest — are not examples of the neat and tidy plant communities I once had faith in. Coniferous forests, like all forests, vary widely in response to local differences in moisture availability, soil

---

Bird Species Commonly Nesting in Coniferous Forests[105]

Hawks and falcons
  Northern goshawk
  Sharp-shinned hawk
  Red-tailed hawk
  Merlin
Gamebirds
  Spruce grouse
Owls
  Great gray owl
  Northern hawk owl
  Great horned owl
  Boreal owl
Woodpeckers
  Northern flicker
Jays and crows
  Common raven
  Gray jay

Chickadees
  Boreal chickadee
Thrushes and kinglets
  Ruby-crowned kinglet
  Gray-cheeked thrush
  Hermit thrush
Waxwings
  Bohemian waxwing
Warblers
  Yellow-rumped warbler
  Blackpoll warbler
Sparrows
  White-throated sparrow
  Dark-eyed junco
Finches
  White-winged crossbill

---

texture, orientation to the sun, permafrost conditions and a host of other factors. Fire in particular can throw a wrench into the tidiest of forest classification systems. Some forests meet textbook expectations and stand clearly apart from their neighbours. Others blend into one another imperceptibly. In the jargon of botany, they may show every degree of floristic intergradation imaginable.

At a gross level, these distinctions among coniferous forests matter little to wide-ranging species of boreal wildlife. The difference between a black spruce and a white spruce is inconsequential to a squirrel looking for cones or a marten looking for squirrels. A spruce or pine will do equally well as an eagle's nest site so long as it can hold the tremendous weight and there are fish nearby. The wide-winged goshawk will fly as gracefully over feathermoss as sphagnum.

But there are more subtle differences among these forest types that can mean life or death for some species, particularly small mammals and passerine birds, which might have relatively confined territories and narrow habitat requirements for reproduction and shelter. The microtopography and insulating qualities of the forest floor — for instance, are there any hummocks or moss? — can be very significant for the winter survival of voles. The presence or absence of a well-defined shrub layer can influence the abundance of juncos, white-crowned sparrows and many other species that build their nests close to the ground. The survival rate of red squirrels may have a lot to do with the spacing of trees, which influences the likelihood of aerial attack by an owl or hawk. To sweep a hand across a coniferous landscape and say that "It's all the same to wildlife" would be to ignore these important ecological relationships.

Like so much else about the taiga shield, many of these relationships are a mystery. I don't know what attracts a hermit thrush to a particular forest near my cabin every year. Is it the tall veteran spruce trees, which provide a commanding perch for its flute-like singing? Or is it the crowded willows below, which offer shelter and protection for a nest site? The fact that I may never know the answers is cause more for celebration than concern.

17

Deciduous Forests: Forests with No Future

Beside my house in Yellowknife is the world's smallest wilderness area. This, at least, is what I lead my neighbours to believe. Smaller than a bowling alley, it contains over 100 trees and shrubs plus uncountable species of native grasses, herbs, mosses and lichens. In winter, ptarmigan dangle from the many willows, snipping off tender, energy-rich buds. In spring, yellow and orange-crowned warblers flit nervously among the birches as they feed on insects or engage in courtship chases. In summer, raspberries and gooseberries swell to red and purple fullness. In fall, the high-bush cranberry and currant shrubs glow red, the birch and poplars orange. All these comings and goings, all this diversity and miraculous order exist in my tiny diorama of life in the subarctic deciduous forest.

What makes this wilderness area all the more satisfying to observe is the fact that I planted just about every stick myself. "Natural landscaping" some call it. Others prefer the term "eco-gardening." By whatever name, the basic goal is the same: to create a unified and self-sustaining landscape modeled after a community of plants found in the wild. The intent is to work with, rather than against, such natural forces as plant succession, birth, death and decay. By relying not on contrivance and whim but on the art and science of nature, the result is a landscape offering a picture of harmony, diversity and subtle beauty.

Although the eventual benefits of a natural landscape are many, getting it up and running is not an easy task, especially when you must start from scratch, as I did. With the promise of a "self-sustaining" (read "zero maintenance") landscape in mind, I set about to transform a bald, dusty roadbed — for that is what it was before my house took shape — into an ecologically credible patch of deciduous woods. The work ranged from labour-intensive to Herculean. Selecting and delivering trees and shrubs from the wild was the most pleasant task. Digging pits for them was the worst, a job made all the more difficult by the need to smash through the cement floor of an old welding shop lying a few centimetres below the surface.

To be true to the ecology of northern deciduous forests, I was obligated to haul in several large burned-out spruce stumps to represent the supposed former forest that went up in flames before the aspen and birch moved in. Beneath the deciduous trees, I transplanted a shrub layer of alders, willows, wild rose, soapberry and several other kinds of light-loving berry bushes. Beneath the shrubs I introduced various kinds of herbs, flowers and groundcover species, such as northern comandra, bunchberry, twinflower and kinnikinick.

To give the developing soil a head start, I stirred in several pickup truck loads of leaf litter carefully gathered from the floor of a mature birch forest. To my pleasant surprise, this material included a liberal mix of seeds and fertile spores, attested to weeks later by the sprouting of several species of mosses, more willows and the ubiquitous fireweed. Judging them to be ecologically correct, I welcomed these unexpected guests to my humble eco-garden.

Into this apparent order, I randomly transplanted several small saplings of white and black spruce. For me, they represent the coniferous vanguards of the forest to come. In theory, the spruce will eventually choke out most of the deciduous species by robbing them of sunlight. It remains to be seen whether I will be around to witness this transformation, which, in a natural setting, would probably take more decades than I have left. In the meantime, with the labour now behind me, I can relax and watch the slow but spectacular process of natural succession.

Thanks to all that digging, hauling and scratching in the dirt, I gained more than a charming little wilderness area and improved muscle tone. I gained a deepened ecological understanding of the forest that served as my model.

Nowhere on the taiga shield are deciduous forests of aspen and balsam poplar very extensive. Paper birch does better at holding its own as a true forest. All of these species occur most commonly as scattered clumps within other kinds of habitats, such as coniferous forests, wetlands and the shallow cracks and dimples on bald rock outcrops. When they do form a forest, the purest deciduous stands occur along well-drained shorelines, on sand or gravel terraces and on warm, south-facing slopes. They may grow to maturity on such sites, producing trees of impressive girth and height.

Sometimes, what looks like a mature deciduous forest from a distance is actually a shrubland dominated by tall willows and alders, with a minor component of deciduous trees sprinkled in. Whether growing as a forest or a shrubland — or one of the many variations in between — deciduous plant communities share several characteristics. They are opportunistic, quickly colonizing areas opened up by fire or windthrow. They are prolific, growing rapidly to assert their access to sunlight. And they are versatile, able to survive a wide range of soil and climate conditions.

Despite these adaptations, deciduous stands are relatively short lived. Their inability to reproduce in their own shade results in what some botanists call "forests with no future." Usually no more than one generation gets a place in the sun before it is overrun by spruce.

I know of mature stands of trembling aspen that appear as groomed and well trimmed as an urban park. The trees have high, mop-topped canopies and nearly branchless trunks. The walking is easy, the atmosphere open, because there are few shrubs covering the forest floor. And aspen saplings — the only hope for regeneration — are virtually nonexistent. The few feeble saplings that poke out of the ground have had their bases debarked or branches snipped clean, having been all but ravaged in winter by snowshoe hare. The lifeless forms of larger saplings, artefacts of less shady times, show signs of being wrenched down and snapped in half by moose browsing on their tender upper branches.

Untouched by hare or moose, scattered unobtrusively among the knotted aspen trunks, are the young spruce. In time, as these trees raise their dark spires up through the deciduous canopy, the aspens will decline in health and vigour, ultimately dropping dead due to lack of light.

In springtime, while most other habitats are resonating with the sound of bird song, the mature deciduous forest can be as empty and still as a locked church. Although a treat to walk through, especially after tunnelling through a young spruce grove or hummock-hopping through the muskeg, some deciduous stands can be among the most impoverished habitats north of 60°, not only in plant species but also wildlife. On the other hand, even when plant diversity is low, they can be among the richest habitats, alive with the flutter and call of nesting birds.

It all depends on the shrub layer. When it is absent or poorly developed, the deciduous forest offers little protective cover for birds and other wildlife. Though many species may pass through to feed on the carpet of berries, it is home to few. When the shrub layer *is* well developed, with thick, crownlike alders and tall willows, the forest's

Common Plant Species of Deciduous Forests[106]

Trees
Paper birch	*Betula papyrifera*
Trembling aspen	*Populus tremuloides*
Balsam poplar	*Populus balsamifera*
White spruce	*Picea glauca*
Black spruce	*Picea mariana*

Tall to medium shrubs
Green alder	*Alnus crispa*
Willow	*Salix bebbiana, S. glauca, S. scouleriana, & others*

Low shrubs and ground cover
Soapberry	*Shepherdia canadensis*
Mountain cranberry	*Vaccinium vitis-idaea*
Highbush cranberry	*Viburnum edule*
Kinnickinick	*Arctostaphylos uva-ursi*
Wild rose	*Rosa acicularis*

Herbs and flowers
Northern comandra	*Geocaulon lividum*
Twinflower	*Linnaea borealis*
Bunchberry	*Cornus canadensis*
One-sided wintergreen	*Pyrola secunda*
Fireweed	*Epilobium angustifolium*

Bird Species Commonly Nesting in Deciduous Forests[107]

Raptors
  Red-tailed hawk
  Merlin
  American kestrel
Gamebirds
  Sharp-tailed grouse
Woodpeckers
  Northern flicker
Flycatchers
  Least flycather
  Alder-flycatcher
Jays and crows
  Common raven
  Gray jay

Thrushes
  Swainson's thrush
Warblers
  Tennessee warbler
  Orange-crowned warbler
  Yellow warbler
Sparrows
  American tree sparrow
  White-crowned sparrow
  Chipping sparrow

morphological diversity is greatly enhanced, allowing different bird species to take up residence at different levels from the forest floor on up.

In the multistoried deciduous forest or shrubland, nesting activities of several bird species are what ornithologists call "geographically stratified." For instance, the Tennessee warbler, orange-crowned warbler and white-crowned sparrow all nest on or near the ground beneath thick shrubbery. The taller shrubs are preferred nesting habitat for the yellow warbler, American tree sparrow and Swainson's thrush. Such common species as the merlin, American robin and northern flicker take to the trees to nest.

Because these breeding birds are separated vertically as well as horizontally, competition for nest sites — and hence feeding territory and singing perches — is kept to a minimum. Consequently, if conditions are right, an amazing number of birds can coexist in a relatively small patch of deciduous trees and shrubs without ruffling each others' feathers.

Such is some of the science of deciduous forests on the taiga shield. Of their magic, I appreciate most how they soften an otherwise rugged landscape dominated by spruce and rock. In springtime, they add a glistening, lime-green thatch to the forest mantle; in fall, a carpet of flames. In winter, their supple, creamy columns take the stiffness out of winter.

Beyond this forest's easily read story of transformation, its berries and its birds, its changing colours and its pleasing shapes, there is something more of special charm to me, something captured in the wind of my modest eco-garden. Beside my home, I have transplanted several trembling aspen trees, not so much for their appearance — they can be quite gangly when young — or for the minimal shade they might offer, but for their sound. Their leaves, the size of quarters, have a thin, flattened stalk, which allows them to quiver in the slightest breeze. The movement of hundreds of leaves from one tree sets off a gentle lisping clatter.

Beside my cabin in the wild, I can stand in the heart of a large aspen grove, with not hundreds but millions of leaves at play in the wind, and hear, quite distinctly, the sound of applause. I like to think of it as the trees giving a standing ovation in praise of the sun.

18

Wetlands: More than Muskeg

The rhythmic red flash of my kayak paddle is all there is to give away my presence to an onshore observer. But then, way out here, who's looking? Maybe a red-winged blackbird riding out the evening breezes astride a favourite cattail reed, or a flicker on top of a half-drowned spruce snag, taking a pause while drilling for dinner. I'm only guessing. All I can see, on all sides and above my head, is a swaying forest of metre-high water horsetails.

With each stroke, my needle-nosed craft swerves slightly right, then left, carving a temporary pathway through the dense growth of wandlike plants. My course is random, guided only by subtle variations in the water depth. I can continue skimming along like this as long as I keep a few centimetres of water below me. The horsetails seem to lift the kayak more than obstruct it, and the going is remarkably easy — getting easier, in fact. I must be near the edge.

Sure enough. I suddenly break into the clear and glide into a still pool of yellow pond lilies set aflame in the low-angled evening sunlight. Completely surrounded by horsetails, this hidden oasis of colour reminds me of a floating Japanese garden: surprising, self-effacing, perfect in every detail.

I get my bearings, then plunge into another thick patch of horsetails. A female green-winged teal rockets up and away from my advancing bow. Had she waited another few seconds, I might have skewered her by mistake. More surprises. Something splashes right beside me, and I look just in time to see a leathery, snakelike tail disappear below the surface — a muskrat, probably on its way to feed on the pond lilies behind me. Then another, much bigger splash, enough to send a spray of water into my face. A line of horsetails shudders and parts, yielding to the muscle power of a quickly retreating northern pike.

That confirms it, there *are* fish in this wetland. I suspected as much when, lowering my kayak at the end of the portage, I spotted a river otter, an adept stalker of fish. It was licking its chops and preening itself on top of an ancient beaver dam fringed with fireweed.

The dam explains why this particular wetland is wet. Placed at some carefully selected hydrological bottleneck — what amazing engineers, those beavers! — it captures water from underground seepage, from a slow-moving stream or perhaps simply from the sky in the form of rain or snowmelt. Stretching almost a kilometre behind a few chewed-up sticks and a bit of mud is a rich wetland ecosystem that chirps, splashes and, for better or worse, literally buzzes with activity throughout the brief northern summer.

With or without the help of beavers, the northern Canadian Shield terrain provides abundant nooks and crannies for water to pool. Repeatedly gouged and plucked by advancing glaciers, the undulating bedrock exerts an unyielding control over the flow of water. Many of today's wetlands started as disconnected hollows and dips in the rock surface offering little or no drainage possibilities for water.

While retreating, the glaciers created more places for water to pool by adding further mayhem to this irregular drainage pattern. As the ice melted, it released bouldery till over some of the region's basement rock. In areas where till was dumped thickly, kettle depressions often formed. Now filled with water, these circular pockmarks in the landscape were created when ice block orphans were left behind by the retreating glacier and later engulfed by a layer of till. The ice block eventually melted, leaving behind an imprint of itself in the gravelly surface.

Which ever way these dimples in the landscape took shape, where drainage is poor or lacking altogether, there today lie wetlands. Some wetlands are flooded all year. Some fill during the flush of spring, then dry up by midsummer. Still others hold water just below the surface, a feature best appreciated when you step in them while wearing a canoe on your head.

Being still or gently flowing, the water in most wetlands carries little oxygen. Starved for oxygen, bacteria and other soil organisms have a tough time breaking down each year's crop of plants and other biological debris. When things pile up faster than they can be decomposed, the result is peat — cold, soggy, smelly, squishy peat — that marvellous mire underlying all wetlands.

My favourite canoe-tripping shoes are entombed in peat, sucked off my feet while I crossed a waterlogged bog. Of all wetlands, bogs are the peatiest. They have claimed other human artefacts, including bulldozers, diesel locomotives and criminals. The latter hailed from

Scandinavia, thrown bound and beaten into a bog 300 years ago for some apparently unforgivable misdeed. So effective is the preserving power of bogs that the menu of their last meal could be determined from the pickled contents of their pickled stomachs.

Bogs come in many shapes and sizes: "raised bogs," which look like giant inverted saucers; "string bogs," whose ribbons of open water and linear hummocks give them a tiger-striped appearance; and "eye bogs," which from the air show a dark, tea-coloured pond for a pupil surrounded by a green iris of floating vegetation. A close encounter with any kind of bog brings one face to face with the common denominator that binds them all: sphagnum moss. Up to 90 percent of a bog's biomass is sphagnum, or peat composed mostly of sphagnum remains.

What distinguishes sphagnum from other mosses is its affinity for acidic growing conditions, a by-product of partially rotted plants, and the peculiar architecture of its leaves. Resembling thin tubes of styrofoam, they consist of large water-absorbing dead cells interwoven with smaller live ones. Thanks to this design, past generations of sphagnum, long dead and buried, can continue to siphon water both upwards and sideways to the growing edge of the sphagnum mat.

Because sphagnum creates conditions fostering its own growth — acidic water and a thickening peat platform on which to root — it can spread, amoebalike, into neighbouring ecosystems, slowly but surely swamping everything in its path in a thick, hummocky carpet of moss. I have seen the bulging waves of gold, red and green sphagnum smother the floor of spruce forests, engulf large granite boulders, swallow whole skeletons of caribou. Only a few acid-tolerant species can make a living on the advancing sphagnum mat: Labrador tea, leatherleaf, mountain cranberry and, where the peat is not too thick, black spruce.

Punctuating the apparent monotony of bog vegetation is a surprising diversity of less conspicuous plants that give expression to the subtle variations in microhabitat created by the hummocky terrain. In drier, open areas, patches of cloudberry often thrive, their pale orange fruits tasting musky and sweet, like baked apples. In wetter, low spots, triangle-stemmed sedges may predominate, including pockets of downy, nodding cotton grass. Where the sphagnum mat reaches out into open water, where it literally quakes when you walk on it, its leading edge often contains a mix of showy flowers such as the purple skullcap, the deep-red marsh fivefinger and the creamy white water arum. And scattered unpredictably and alone through the bog, as if planted by mischievous fairies, are its diminutive orchids, including the fragrant ladies'-tresses orchid, so-called for its spiralling floral pleats resembling fancy braids in a woman's hair.

Fens are another common wetland in shield country. Like bogs, they are underlain by thick peat. But a knowing eye can spot the difference between these two kinds of wetlands in an instant. Bogs are virtually choked off from flowing water, the main reason for their acidic, nutrient-poor condition. Fens, on the other hand, are bathed in a shallow seepage of water that, though sluggish, is sufficient to flush out acids and maintain a relatively healthy supply of nutrients and oxygen. Sphagnum doesn't have a chance in this more alkaline environment. Instead, the peat is covered with a dense carpet of sedges, the distinctive vegetation cover of northern fens.

Fens often merge into marshes in shallow, sheltered areas along rivers and lakes, adding a soft green fringe to an otherwise abrupt and rocky shoreline. Subarctic marshes are dominated by plants that got the tail end of botanical nomenclature: cattails, horsetails and mare's tails. Bur reed, duckweed and red dock are other typical plants that can put up with the periodic flooding and drying that characterize life in a marsh.

Beaver ponds, bogs, fens and marshes — in some places you will discover these northern wetlands as small, discrete and easily recognizable packages, classics in their own right. Elsewhere you'll find that they coalesce with each other, forming vast expanses of waterlogged muskeg. From the air, muskeg presents itself as a glinting waterland, with the sun reflecting off thousands of small puddles and ponds held within a spongy maze of hummocks. Though muskeg may be pleasing to behold from above, try a June walk through such terrain without getting a soaker, getting lost or, worse, losing a few litres of blood to mosquitoes and blackflies.

The size and character of northern wetlands are as varied as the sentiments they arouse in different people: dank, unpleasant places infested with bugs, or havens of life for a lush assemblage of plants, birds and mammals? Evil-smelling backwaters with the consistency of quicksand, or pleasing oases of rich and varied colours and textures? Wetlands can be all these things, and much more. For me, their ecological virtues far outweigh the occasional hardships they may impose upon our own frail species. Wetlands act as giant sponges, soaking up rain and snowmelt, then slowly releasing much-needed water to drier habitats around them. Like kidneys, wetland plants act as filters absorbing or breaking down errant pollutants that may enter the water. Their relatively warm, shallow waters and rich supply of organic matter provide the basis for a level of biological productivity unmatched in other northern ecosystems. For instance, over the span of less than two months, a cattail marsh can produce a dense stand of sturdy reeds over two metres tall. Such luxuriant growth, in turn, supports complex aquatic food chains, provides rearing areas for developing fish fry and offers essential food and cover for muskrats and beavers as well as nesting or moulting waterfowl.

Beyond their ecological values, wetlands offer unique aesthetic opportunities to experience northern nature at its concentrated best. The observer initiated into the wonders of northern wetlands can constantly discover new sources of charm and beauty

Common Plant Species of Wetlands[108]

Lower plants
    Water horsetail   *Equisetum fluviatile*
    Marsh horsetail   *Equisetum palustre*

Submerged and emergent plants
    Cattail   *Typha latifolia*
    Bur reed   *Sparganium angustifolium, S. minimum*
    Pondweed   *Potamogeton gramineus, P. richardsonii*
    Water arum   *Calla palustris*
    Duckweed   *Lemna* spp.
    Water smartweed   *Polygonum amphibium*
    Yellow pond lily   *Nuphar variegatum*
    Water milfoil   *Myriophyllum exalbescens, M. verticillatum*
    Buckbean   *Menyanthes trifoliata*
    Mare's tail   *Hippuris vulgaris*

Sedges and grasses
    Marsh sedge   *Carex aquatilis, C. rostrata,* & others
    Cotton grass   *Eriphorum angustifolium*
    Bulrush   *Scirpus caespitosus*

Mosses
    Bog moss   *Sphagnum* spp.; *Drepanocladus* spp. *Aulacomnium* spp.
    Fen moss   *Tomenthypum nitens*

Trees
    Black spruce   *Picea mariana*
    Tamarack   *Larix laricina*
    Paper birch   *Betula papyrifera*

Tall to medium shrubs
    Willow   *Salix bebbiana, S. myrtillifolia, S. planifolia* & others
    Dwarf birch   *Betula glandulosa*

Low shrubs
    Sweet gale   *Myrica gale*
    Leatherleaf   *Chamaedaphne calyculata*
    Bog laurel   *Kalmia polifolia*
    Labrador tea   *Ledum groenlandicum*
    Small bog cranberry   *Oxycoccus microcarpus*

Herbs and flowers
    Northern bog orchid   *Habenaria obtusata*
    Ladies'-tresses orchid   *Spiranthes romanzoffiana*
    Red dock   *Rumex occidentalis*
    Bog star   *Parnassia palustris*
    Marsh fivefinger   *Potentilla palustris*
    Round-leaved sundew   *Drosera rotundifolia*
    Short-stemmed raspberry   *Rubus acaulis*
    Cloudberry   *Rubus chamaemorus*
    Water parsnip   *Sium suave*
    Labrador lousewort   *Pedicularis labradorica*
    Skullcap   *Scutellaria galericulata*
    Mastodon flower   *Senecio congestus*

Bird Species Commonly Nesting in Wetlands[109]

Grebes
    Red-necked grebe
    Horned grebe
Ducks
    Green-winged teal
    Mallard
    Northern pintail
    Blue-winged teal
    Northern shoveller
    American wigeon
    Ring-necked duck
Rails and coots
    Sora rail
    American coot
Wading birds
    Sandhill crane
Shorebirds
    Semipalmated plover
    Killdeer
    Lesser yellowlegs
    Least sandpiper
    Common snipe
    Red-necked phalarope
Hawks
    Northern harrier
Owls
    Short-eared owl

Flycatchers
    Alder flycatcher
Swallows
    Tree swallow
Thrushes
    Swainson's thrush
    American robin
Warblers
    Tennessee warbler
    Palm warbler
    Northern waterthrush
    Wilson's warbler
    Yellow warbler
Sparrows
    American tree sparrow
    Chipping sparrow
    Savannah sparrow
    Fox sparrow
    Song sparrow
    Lincoln's sparrow
    Swamp sparrow
    White-crowned sparrow
Blackbirds
    Red-winged blackbird
    Rusty blackbird

in familiar landscapes previously written off as "nothing but muskeg and sloughs." On display in taiga wetlands are a host of miraculous works of art — like the green-winged teal scared up by my kayak.

Drifting now in the middle of the beaver pond, I rest my elbows on my paddle, eyes closed, and listen to the evening's entertainment. From way above my head comes a ghost-like tremolo noise straight from the soundtrack of an old Tarzan movie: "who-who-who-who-who-who-who-who-who." An impassioned common snipe is putting on an aerial show for his mate hidden among the sedges. As he plummets towards her, his spread tail feathers buzz rhythmically in response to bursts of air rushing out from behind his rapidly beating wings. Circling around me is a flock of tree swallows engaged in busy chatter as they scoop mouthfuls of newly hatched mayflies and mosquitoes from the air. I duck instinctively as the swallows circle closer to me, perhaps curious, perhaps seeing me as mosquito bait.

The sudden movement of such a large mammal as myself triggers a new sound: a series of rapid-fire "tew" notes emitted from the beak of an obviously ticked-off lesser yellowlegs. After a few long minutes of remorseless scolding, the bird is finally convinced of my harmlessness, calms down and returns to its treetop sentry post. Then from behind me, a peculiar squeaking, mewing noise, followed by silence, the sound of chewing, a large splash, then more squeaks. What in the world? I've got to open my eyes for this one.

I turn around to discover that the stern of my kayak is now less than a metre away from a large beaver lodge. How long have I been drifting? There must be a bunch of young beaver kits in there, probably enjoying a pre-bedtime snack until a doting parent sensed my presence and left the lodge to investigate. "Slap!" A beaver tail meets the water just in front of my boat. The message is clear: scram.

I raise the sharp blade of my paddle, pausing for a moment to savour the stillness. The water shimmers like liquid mercury and I am hesitant to rupture its perfect film. The kayak comes alive after a few strokes as I glide towards the beaver dam.

Tall spruce and tamarack, well back from shore, send a candelabra of dark blue shadows across a narrow ring of willows and shrub birch, across a sedge meadow dotted with islands of Labrador tea, across the horsetail marsh I paddled through earlier and out into the middle of the pond. Along the opposite shore, a rock outcrop rising steeply out of the water is bathed in warm orange light. Lit more from below than above, the solitary jack pines that cling to its cracks and crevices seem to glow green from within. The sky above the rock wall is awash with all shades of purple and blue and is capped with a light dusting of wind-shorn clouds the colour of amber. Above the clouds hangs a smoky half moon, mine alone this moment.

19

Rivers and Lakes: Young, Wild and Free

Out of the trees, into the trees. Out of the trees, into the trees. Up, down. Up, down. Each hill we climb is bald and treeless, as if shorn by the wind. Depressed shrubs and a spongy carpet of lichen and moss tell us that we are on true tundra. Walking downslope, we re-enter the taiga forest, complete with dense-packed spruce and fallen snags to hinder our progress through the wilderness. Here, near the tree line east of Great Slave Lake, we are walking between worlds, following a compass bearing that is supposed to take us directly overland to Parry Falls, the Northwest Territories' third-largest waterfall.

Soon after setting out from Toura Lake, one of a series of water bodies along Pike's Portage, we realize that nobody walks directly overland in this country. There are too many small lakes and wetlands in the way to call any route direct. On the map, our route one way is 8 kilometres. After a morning's walk, we have travelled perhaps 12. While undulating over countless ridges and skirting as many lakes, we get lost.

"Your compass lies! How could the falls be over that ridge? Look at that little bay with the hill behind it. From the top of that hill we should be able to look right down on the falls. . . ." So the discussion goes, as our party of four landlocked canoe trippers navigates by whatever means suits for the moment — by compass if it feels right, by lake shape if it doesn't. We opt for the hilltop view. Where the falls should be, we discover a river of stunted black spruce.

Before the discussion is allowed to go further downhill, we all sit down for a "gorp" break — good old raisins and peanuts — and gather our collective wits. Between munches, we hear only gentle gusts of wind whispering through the spruce below us and an occasional lisping clatter made by the wings of hovering dragonflies.

With renewed mental reserves, we attack the well-worn Lockhart River map sheet, trying to wrest some sense out of it by concentrated stares and serious guesswork. The wind dies suddenly, snuffed out by the searing midday sun. One by one, we look up from the map in response to a new sound. It is low, pulsing, barely audible, yet powerful — as if the earth itself were chanting some kind of mantra. Like a mine detonation deep underground, we hear it more with our stomachs and feet than our ears.

Then everything makes sense. "Can you feel it?" I ask. "It's the falls!" The map is hastily folded, shoved into a pack and forgotten. The sound leads us to the crest of a neighbouring hill, where it instantly becomes a sustained roar — the sound a good-sized river makes when it is throttled through a narrow bedrock canyon and freefalls 40 metres to a seething whirlpool below. Our cheers and laughter are muted by the thunder of Parry Falls as we dance along a cliff edge, high above the spray.

Situated halfway down the 30-kilometre Lockhart River, Parry Falls is one of hundreds of major and minor obstructions that introduce chaos into the otherwise orderly flow of water downhill. The Lockhart is among the many rivers in this region that, from a drainage perspective, have been described variously as disrupted, disorganized, interrupted, haphazard — even deranged. You would have to be a little deranged to try paddling the full length of some of the rivers that spill over the taiga shield.

In 1833, Captain George Back tried paddling down the Lockhart River on his way back from the barren lands. Fortunately, after witnessing the sudden transformation of one of their birch-bark canoes into matchsticks, he and his crew had enough sense to get out and walk.

> At first we walked with tolerable speed over the broken rocks, and through the intersecting gullies; but the kind of ladder exercise which this imposed taxed the muscles so severely, that the strongest was fain to slack his pace, as the same interruptions and impediments multiplied upon us.[110]

About 150 years later, an intrepid middle-aged couple from Germany attempted the same route mainly for the fun of it. In their trip journal, Walter Frueling and his wife, Renata, admit up front that they are "no whitewater specialists."[111] They exercised extreme caution. They did an awful lot of river scouting, canoe lining and portaging. This was wise. In the *International Scale of Whitewater Difficulty*, much of the Lockhart River deserves at least a class 5 ranking:

Extremely difficult, long and very violent rapids with highly congested routes which should always be scouted from shore. Rescue conditions are difficult, and there is significant hazard to life in the event of a mishap. Ability to Eskimo roll is essential for boaters in kayaks and decked canoes.[112]

With a keen eye for detail, the Fruelings described the more outstanding stretches of what they simply called "bad river." Their many careful drawings of rapids, falls, eddies and "funny cross waves" give the impression of anatomical sketches of a slightly inflamed small intestine complete with multiple constrictions, random swellings, ungraceful curves and edges that are generally convoluted.

Why, along the rivers of shield country, are there so many "interruptions and impediments," as described by Back and sketched by the Fruelings? And why so many lakes?

The waxing and waning of glaciers had a lot to do with it. On their many advances over the shield, glaciers removed much of the soil and thick sedimentary deposits (for instance, limestone), which in many other landscapes have a subduing effect on drainage patterns. Throughout much of shield country, bald bedrock — with its many faults, dykes, joints and fractures — exerts an unyielding control over which way the water flows.

While advancing, the glaciers also gouged and plucked away at the rocks, creating numerous disconnected depressions now filled with water. The three "great lakes" of the taiga shield — Athabasca, Great Slave and Great Bear — were formed this way. Bearing down out of the northeast at around three billion tons per square kilometre, the ice began scooping in earnest once it reached the geological divide between the Canadian Shield and the softer sedimentary rocks of the interior plains. In the wake of the ice, three massive basins were left behind that today plumb up to 600 metres in depth.

While retreating, the glaciers added further mayhem to the irregular drainage patterns that would follow. As the ice melted away, it released unimaginable torrents of debris, which is now strung across the landscape in the form of corrugated piles of boulder-strewn till, huge esker trains 100 or more kilometres long, outwash aprons of crudely sorted sands and gravels formed by meltwater streams and raised beach ridges formed along the shores of phantom proglacial lakes. Conforming more to the contours of rotting ice than the shield landscape below, this material was laid down largely at random. As a result, many pre-glacial drainage channels became choked or dammed outright by glacial debris.

Ever since the last chunk of glacial ice melted away, the streams and rivers of shield country have been seeking a new equilibrium. They do this by carving down, ever so gradually, through the bedrock and glacial debris, creating new drainage channels between lakes, removing obstructions that make waterfalls and rapids and straightening out erratic courses. This process will take a liberal allowance of time. Relative to the

Water follows the path of least resistance over bald Precambrian bedrock. *(John Poirier)*

hydrological history of North America, the rivers and lakes of the northwestern Canadian Shield are quite young — "juvenile" in fact, as they say in the trade.

On the way to maturation, the pattern of rivers and lakes — like the water they carry — will constantly change. It may happen suddenly, as in the highwater breaching of a morainal dam at the end of a lake, or it may happen slowly, imperceptibly, like the incising and widening of a bedrock gorge.

Besides the erosive action of flowing water, there is another force at play that will have a subtle but ultimately profound influence on northern drainage patterns, particularly lakes. Glaciers again are largely responsible, in this case due to their *absence*. When present, the ice pushed this part of the continent hundreds of metres into the plastic molten mantle below (much as a floating cork will sink deeper if loaded with spare change). Unburdened of this load a little over 8,000 years ago, the shield rock is now bouncing back, in some places by a rate as much as half a centimetre per year. As the land rebounds, water-filled dimples and veins in its surface will become shallower and drain more quickly. Gradually, the land's ability to catch and hold water will decline.

Such processes lead Yellowknife geologist John Brophy to conclude that, no matter how many lakes we may have now, they are all "geologically ephemeral entities." He cites the shrinking of the once great Glacial Lake McConnell into three separate smaller lakes as proof that the lakes of shield country are on their way out.

> Fifty thousand years from now, a mere burp in the 4,600-million-year history of the earth, the processes of rebounding and erosion will have erased our legacy of lakes.[113]

In the meantime, the rapids will be run, the lures will be cast, the boat trips will be planned and launched — rain or shine. For the brief span of my life, I will assume that the rivers and lakes of shield country are here to stay.

While we, these northern waters and I, coexist in this place and time, I will seek to become more intimate with their ways and wonders. Paddling the air while my canoe hangs over the edge of a boiling haystack brings quick and long-lasting insights. Watching a depth sounder chart invisible crags and valleys hundreds of metres below a moving boat plumbs the depths of my imagination. Seeing an osprey plummet into a still river, then take wing with a silver, wriggling fish in its talons makes me ponder and praise the rhythms of life these waters support.

One of my favourite, and increasingly addictive, means of learning about the aquatic world is a plunge beneath the surface equipped with mask, fins and snorkel. The barrenness of some underwater scenes that greet me testifies to the low biological productivity of all subarctic systems: bald, rocky shorelines that blend into bald, rocky lake bottoms showing no shred of green; long stretches of river bottom covered in bare muck, interrupted only here and there with tiny clumps of grasslike quillwort. Yet, other scenes present a picture of aquatic life in teeming abundance.

Swimming, eyes down, through a thick stand of water horsetail, I revel in the warm, shallow waters along the margin of River Lake. I am a lumbering behemoth floating through the upper canopy of this forest in miniature. I am an infrequent visitor from the dry land. Yet I am very much at home here, perhaps because of my orientation to the stars at birth: Pisces, the fish.

With light flutters of my fins, I slowly cruise through dense, swirling clouds of minuscule water shrimp, daphnia and other invertebrates, which provide the anchor at the bottom of many aquatic food chains. Sailing through these clouds like tiny six-legged blimps are roving bands of red water mites, so-called underwater wolves of the invertebrate world. My shadow scares up a school of darting minnows, probably lake trout. How many of these, I wonder, will survive to grow into gold-dappled adults that retire to the cool depths in spring, spawn in rocky shallows in autumn and challenge an angler's skills below an ice hole in winter?

Drifting over a clear patch of gravel and fine silt, I suddenly stop in response to some slight motion on the bottom. It may have been the bobbing of these snail shells that caught my eye or the jerky march of that caddis-fly larva encased in its stick-built mobile home. A small puff of silt erupts from the bottom, and I catch sight of a retreating ninespine stickleback. It is only a few centimetres long and perfectly camouflaged against the pale grey-green background. No wonder I didn't see it right away.

I move into deeper water towards a swaying jungle of metre-high pondweed. A pair of walleye fingerlings races up to the surface as I descend. My weight belt keeps me at a neutral buoyancy and I can skim effortlessly along just centimetres from the bottom. Well into the pondweed, I come to an abrupt halt, suddenly finding myself nose to nose with a mid-size northern pike. It hangs absolutely still except for its lacy pectoral fins, which quiver gently like two free-floating leaves.

Here, a tail flick away from my mask, is the consummate predator of many boreal lakes, a fish whose mastery at stillness — for both offence and defence — is matched by its mastery of speed. The mighty "jack" is said to have an insatiable appetite for any animal it can engulf with its wide, well-toothed jaws. Newly born pike fry dine first on aquatic invertebrates but soon become cannibalistic, setting on other fry of their own kind that are almost equal in size. As adults, fish, frogs, ducklings, even small muskrats are all fair game. I once saw a pike leap out of the water and onto the shore in pursuit of a pair of colourful socks — while still on my wife's feet!

Though I'm too big for this pike to eat, I edge respectfully backwards, then shoot up to the surface, my breath gone. A lone common loon, half-submerged, swims towards me, perhaps looking for some of the fish I stirred up. We call back and forth to each other until it dives. Might I see it swim past me underwater? Not this time.

To beat the chill of my underwater tour, I stretch out on a smooth outcrop of ice-polished granite and soak in the rays of a mid-afternoon sun. A gentle onshore breeze keeps the bugs off me and breaks the water into a constellation of exploding stars. My eyes are drawn from this hypnotic view by the electric rattle of a belted kingfisher.

It appears out of nowhere, streaks towards a small, shallow bay downshore, then returns to a nearby willow branch hanging over the water. Paddling the air with strong, undulating bursts, it repeats this flight path several times. On its fourth round, the kingfisher does a quick zigzag over the bay, then stops in mid-air, hovering tensely over

---

Common Plant Species along the Shores of Rivers and Lakes[114]

Lower plants
    Water horsetail   *Equisetum fluviatile*
Submerged and emergent plants
    Quillwort   *Isoetes muricata*
    Pondweed   *Potamogeton gramineus, P. richardsonii*
    Water smartweed   *Polygonum amphibium*
    Yellow pond lily   *Nuphar variegatum*
    Water milfoil   *Myriophyllum exalbescens, M. verticillatum*
    Buckbean   *Menyanthes trifoliata*
Trees
    White spruce   *Picea glauca*
    Paper birch   *Betula papyrifera*
    Balsam poplar   *Populus balsamifera*
Tall to medium shrubs
    Willow   *Salix serissima, S. arbusculoides, S. alaxensis,* & others
    Grey or hoary alder   *Alnus incana*
Low shrubs
    Sweet gale   *Myrica gale*
    Shrubby cinquefoil   *Potentilla fruticosa*
Herbs and flowers
    Wild onion   *Allium schoenoprasum*
    Red dock   *Rumex occidentalis*
    Bog star   *Parnassia palustris*
    Marsh fivefinger   *Potentilla palustris*
    River beauty   *Epilobium latifolium*
    Water parsnip   *Sium suave*
    Skullcap   *Scutellaria galericulata*
    Indian paintbrush   *Castilleja raupii*
    Northern yarrow   *Achillea nigrescens*
    Bur marigold   *Bidens cernua*
    Sweet coltsfoot   *Petasites sagittatus*
    Groundsel   *Senecio streptanthifolius*

a prospective meal. It takes aim, pulls in all flaps and dives headlong into the water, straight as an arrow. Before the spray settles, the bird emerges with a small fish in its beak. The whole capture sequence lasts about a third of a second. I grab my binoculars (never far from my side) and watch the kingfisher return to its lakeside perch. The fish is still wriggling, but not for long. Several swift bashes against the willow branch stun the fish enough for the bird to start picking away at it and finish it off.

The warm, shallow waters fringing many boreal lakes and rivers are, relatively speaking, a hotbed of biological activity. The rich life they support below the surface draws much activity from above. I consider the airspace along the shore to be a multilevel flight corridor, travelled frequently not only by kingfishers but also bald eagles, spotted sandpipers, mew gulls and arctic terns, to name a few. After my vigil with the kingfisher, all of these species flew past me in the space of half an hour, all hunting the same productive waters in their own unique way.

Shorelines hold a special fascination for me also because of the land plants that cling to them. There are few more challenging places to take root. Pounding waves and storm surges can eat away hunks of peaty shoreline over the course of one summer squall. Huge pans of grinding ice ride up and over the shore in springtime. Water levels fluctuate

---

Bird Species Commonly Nesting by Large Rivers and Lakes[115]

Loons and grebes
- Common loon
- Pacific loon
- Red-throated loon
- Red-necked grebe
- Horned grebe

Ducks
- Canvasback
- Greater scaup
- Lesser scaup
- Surf scoter
- White-winged scoter
- Bufflehead
- Common merganser
- Red-breasted merganser

Hawks and eagles
- Osprey
- Bald eagle

Shorebirds
- Semipalmated plover
- Spotted sandpiper

Gulls and terns
- Bonaparte's gull
- Mew gull
- California gull
- Herring gull
- Caspian tern
- Arctic tern

Kingfishers
- Belted kingfisher

Flycatchers
- Eastern phoebe

Swallows
- American tree swallow
- Barn swallow

Warblers
- Northern waterthrush

Sparrows
- Song sparrow

widely from season to season and year to year. Among the plant species that can handle such shoreline rigours are some of my favourites: the wild onion, groundsel, sweet gale and water parsnip.

Ironically, just beyond this extremely dynamic environment are some of the oldest, most stable living plants in shield country — magnificent stands of riparian white spruce. The well-drained, porous soils found near the edge of many lakes and rivers offer an optimal growing bed for this species. Here, the odds against destruction by fire are lowered by relatively moist conditions, cooler air and banks that often slope steeply into the water. Throughout this region, these conditions result in isolated ribbons of towering veteran spruce, bounded by much younger forests on one side and by water on the other.

Nothing beats spruce for musical resonance. When sliced thin, kiln-dried and cured in fine lacquers, it is *the* wood of choice for the soundboard of most string instruments — from slide guitars to Stradivarius violins. Similarly, when spruce trees stand tall beside a lake lined up branch to branch, nothing resonates better to the music of shield country.

This thought occurred to me one June evening while lying half awake in a tent on the north shore of Great Slave Lake near Rae. For some time, I had been listening to the violent play of waves crashing on bedrock. I gradually became aware that above these vibrations was a softer, similar rhythm, slightly out of synch with the waves. It was the spruce wall behind me sending echoes back towards the lake. The effect was mesmerizing, alluring. I felt called out of my tent.

It was almost midnight, yet a sliver of sunlight still shone on the turbid, caramel-coloured waters of the North Arm. I discovered that besides echoes, the trees were extending deep purple shadows almost half a kilometre out onto the lake. While breathing in this dreamlike scene, my drowsiness spontaneously gave way to a vivid clarity of mind. My spirit leaped like a lake trout rising in autumn. For this finely woven tapestry of sound and light dancing on the waters, I was moved to give thanks.

20

The Tree Line: Life on the Frontier

"I don't believe in the tree line, but it just ended, right down there!" I shout these words not because of the engine noise but because of my excitement. The hunter from Iowa sitting behind me smiles and nods without taking his eyes away from the plane window. He is watching for caribou. I am watching the trees. Over the space of a kilometre, they just vanished. We have crossed the magical boundary line between taiga and tundra.

I look again, this time with my binoculars. Aha! Snags, lots of them, standing guard defiantly over an empty landscape. And on the ground, grey spars lie crisscrossed like spilled matchsticks. That wasn't the tree line at all but the work of an intense fire, which left behind a clean-edged swath of destruction through the forest. I figured something was fishy. Besides, we're only a half hour out of Yellowknife.

After a few more kilometres, spruce trees pick up again, as big as ever. They are widely spaced on a yellowish green carpet of lichen. Weaving around the wooded uplands are string bogs, ribbonlike fens and countless ponds and lakes. This is lichen-woodland country. We have a way to go yet. The trees suddenly vanish again, this time under a cloud.

Half an hour later, the clouds part. It's as if the landscape had spun itself a misty cocoon and emerged totally transformed. My initial impression is that, for the first time in around two billion years, the land is once again molten. It seems to boil with colour. Rivulets of red and copper ooze like blood around cells of exposed bedrock. It's September and the tundra shrubs are wearing their finest colours. A mossy blanket of lime green and orange polygons accents the lowest ground. Cutting through these patterns are trains of eskers, snakelike mounds of sand and gravel that once filled channels on, in or under the retreating glacier.

Splashes of boreal green hug the eskers' shoulders and crowd along the lakeshores and many streams. The last of the spruce grow in small clumps or alone. They grow in narrow fingers thrust tentatively out into the barrenlands. They grow wherever they find shelter from the worst blasts of winter wind. These are the sentinels of the forested frontier I have come to explore.

Our final descent onto the waters of Humpy Lake brings us alongside a particularly large and long esker — this is one big hump! Before the waves started lapping away at its middle 8,500 years ago, this esker may have stretched from one side of the lake to the other, a distance of 5 kilometres. Away from the lake, its longest remaining piece is 20 kilometres. Many other eskers wind unbroken for twice that length. They make great avenues for migrating caribou, offering smooth, dry footing, few bugs and a clear vantage for spotting predators.

"There's a buck!" It's the Iowa hunter's turn. While still airborne, we spot a bull caribou (apparently it's "buck" south of the 49th parallel) looking *down* on us from the top of the esker. It calmly watches us land and then disappears down the far side. Two other caribou trot away along the crest as our Twin Otter slides to a halt on the beach. While we wait for the motors to die, I quickly check the topo map for signs of green. It tells me that we have landed on the outer limits of the tree line.

We stopped here because this is where we happened to see the most caribou. Why the trees stopped here is another matter. For years, meteorologists, botanists and boreal ecologists have challenged each other with possible explanations. The debate still goes on and may yet for some time.

The standard approach to figuring out what's causing vegetation patterns in any landscape is to make a bunch of map overlays showing bedrock geology, surficial geology, soils, topography and drainage, superimpose them over a vegetation map and then start looking for correlations. Scientists investigating the tree line found this exercise very fruitful in discovering what was *not* causing it.

They followed the tree line all the way from the Mackenzie Delta to the Labrador coast and found no positive correlation. Certainly the Canadian Shield had nothing to do with it. The rock was basically identical on both sides of the trees. Soils did show some differences in their texture, chemistry and permafrost conditions, but the scientists didn't know if they were looking at a cause or effect, since the presence of trees can influence all of these factors. They discovered no great ups or downs in the topography that might provide a clue. Only when they added the climate overlay did it begin to make sense.

Oddly enough, temperatures in summer, not winter, may be the crucial factor in determining where the tree line rests. In July, at the arctic air mass's southern boundary, temperatures average 10°C. Overlay this temperature boundary, or isotherm, onto a vegetation map and, *voilà*, correlation. The tree line sits pretty well right underneath it.

On the ground, what actually stops the trees from advancing beyond this isotherm? A popular explanation is that spruce trees need an average July temperature of 10°C or more for sexual reproduction. Anything less and successful fertilization, seed germination and seedling growth become impossible. What it boils down to is that the growing season is just too short to cram in all these activities.

On the other hand, spruce trees that reproduce exclusively by sending sprouts up from the roots — by layering — can carry on beyond this critical threshold, growing in relative isolation out on the tundra. But ultimately they too reach their limit, held at bay by the icy winds of winter.

The tree line is a unique ecological phenomenon that can be as confusing as it is fascinating. Not surprisingly, it goes by many other names: the "continental arctic tree line," the "approximate northern limit of trees," the "northern forest border," the "northern forest-tundra ecotone" and the "far northern tree line." I could list several more. All these names, with varying degrees of success, point to the same thing. My reason for favouring the term "tree line" is brevity.

Grade school teaches us — or at least it taught me — that somewhere in the Far North was a place you could go and gaze upon a clearly defined wall of trees, stretching from horizon to horizon, with absolutely nothing at your back but tundra. There are some places where you actually can do this. Near Ennadai Lake, for instance, off the northeastern corner of Saskatchewan, the tree line is a clearly visible entity — as in my boyhood dreams. Here a dense forest of spruce trees, some of them 10 metres high, suddenly comes to a dead end and tundra takes over — for the next 1,000 kilometres (after that it's salt water). Such sites are the exception.

The norm is not a line that you can reach out and touch. It is rather a *zone* of transition where spruce trees intermittently come and go, then gradually fade out completely. For much of the taiga shield, this zone lies between the great lichen-woodlands to the west and barren tundra to the east. It averages about 50 kilometres in width — this would make a pretty fat line on a map.

Local variations in topography, climate and fire history create unexplained wobbles in this transition zone. In some places, it is over 100 kilometres wide. In others, this distance is telescoped down to a mere sliver of land. In all of shield country, the shortest and easiest overland hop from taiga to tundra is made from the end of the East Arm of Great Slave Lake. People have been making this trip for centuries.

The original Chipewyan who lived in the East Arm area enjoyed the best of both worlds. Thanks to an unusual pinch in the tree line, they had easy access to two relatively distinct reserves of wildlife. On the floor of their tents, one might just as easily have found a hide from a muskox as a black bear.

In a journey done in under two days — with a good tail wind — the Chipewyan followed a well-worn trail and chain of small lakes from the closed spruce forests on the shores of Great Slave to the barren lands bordering Artillery Lake. In 1889, an adventurer named Warburton Pike followed in their footsteps once or twice and somehow got the route named after himself: Pike's Portage. He admitted later in his journal that before he arrived, the Chipewyan had been using this convenient route across the tree line "from time immemorial."[116]

It was the same for the Dogrib and Yellowknife peoples. They had their own favourite routes across the tree line: up the Yellowknife, Snare and Thelon rivers and down the Coppermine. These were longer journeys but equally important: they allowed them to diversify their options for getting meat. Archaeological evidence suggests that Dene hunting parties travelled at least as far east as Baker Lake to pursue muskox and caribou on the tundra. In contrast, the Inuit stayed pretty much out of the woods.

In this sense, the tree line was a cultural as well as ecological boundary. It still is, to the point of being highly politicized. Since the 1970s, the tree line has been the geographic centrepiece for all political debate on settling east-west land claim issues, splitting the Northwest Territories in two and developing a home-grown constitution for the common good of all peoples — on both sides of the trees.

While I sit in the sun on a lookout above our Humpy Lake camp, the tree line's political ramifications are far from my mind. What interests me is the pocket of trees beside me — if you can call them that. The taller ones look like crumpled witches' brooms that have crash landed without their riders. Most of the trees have clumps of small, gnarly branches along their trunks on only one side. They've been "flagged." Their growing buds were killed off on the windward side by the abrasive effects of ice and snow travelling at speeds up to 120 kilometres an hour.

Spreading below the trees is a dense cushion of spruce about 60 centimetres high. This is *krummholz*, a German word meaning "crooked wood." Though crooked indeed, the branches appear much healthier than on their taller neighbours, having played it safe through many winters under a protective blanket of snow.

I get down on my elbows and knees and stick my nose into the krummholz cushion to see what lies below. The ground is choked with familiar low shrubs that find shelter among the spruce branches: mountain cranberry, Labrador tea and bog laurel. The scene strikes me as a kind of mini taiga forest unto itself. The spruce stems are remarkably thick, about the diameter of my arm. Many grow up independently through the shrubs. These aren't branches at all but dwarf spruce trees, with lives of their own. Through layering, they probably sprouted as genetic clones from the low, spreading branches of some long-dead ancestor.

This isolated patch of krummholz didn't just land here out of the sky. Long ago, at least one spruce seed must have blown to this high spot, germinated and taken root. Out here on the tundra, this kind of thing doesn't happen anymore — the conditions are too harsh. There are a surprising number of cones in some of the taller spruce but their seeds are infertile. No sex on the tundra for these trees.[117]

Through layering, patches of spruce like this may have hung on for decades, maybe centuries. They likely represent relics from warmer times when the forest extended farther north. As long as the climate remained stable, fires probably came and went through this forest, cones released fertile seeds onto the ash, seedlings sprouted, the forest held its ground. But a climatic cooling trend may have brought summer temperatures below the critical threshold where sexual reproduction is possible. At that point, the next big fire would have wiped out the forest's northern fringe, pushing the tree line back to some new climatic equilibrium. The tundra carpet would have moved south, parting here and there for isolated clumps of surviving trees that may have had a knack for layering.

I can only speculate. My long communion with the krummholz has left my elbows and knees sore. As I lie down on a mattress of crowberry, I find myself wishing that these spruce could tell their own story.

A pair of Pacific loons flies across my view of the sky. In a few days, they may be dining on herring off the Baja coast of California. I dig into my pack for binoculars and follow the loons out of sight. See you next year.

I scan the lakeshore for shorebirds. I scan the trees for tree sparrows. I listen for the horned lark. Nothing. I do this mostly for nostalgia's sake to keep alive my memories of summer at the tree line.

In September, the land is largely birdless. Most migrants have long gone. Among them are a few bird species that feel equally at home in the taiga or tundra: the red-breasted merganser, the northern pintail and the American tree sparrow. For them, the tree line has little, if any, significance.

These species are unusual. Most birds, from both sides, recognize the tree line as an inviolable habitat threshold. What is most amazing is the number of birds whose northern limit corresponds exactly with the tree line (however exact that can be). For instance, although there are thousands of small ponds out on the tundra, you will listen in vain for the peep of a green-winged teal or the quack of a mallard. Except for the pintail, no dabblers cross that line. Conversely, the only time Lapland longspurs, semi-palmated sandpipers, golden plovers and other strict tundra breeders cross the tree line is during migration.

There are a few birds for whom the transition zone between taiga and tundra is preferred nesting habitat, namely, the gray-cheeked thrush, Harris's sparrow and northern shrike. These are the true ecotone dwellers that, like the Chipewyan of the East Arm, make the best of both worlds.

Mammals show similar patterns. For a minority of species, the shape of the tree line is not reflected in their overall range. Moose, wolves, wolverine and of course caribou cross freely from taiga to tundra or vice versa. Red-backed voles have what it takes to adapt to both worlds. Most other small mammals are bound by the tree line to one side or the other. For instance, collared and brown lemmings shun the trees as earnestly as pygmy shrews and deer mice shun the tundra.

For many larger species that represent the animal emblems of all that is boreal — the red squirrel, the porcupine and the beaver — their ranges come to an abrupt halt at the tree line. Out on the tundra, what is there of substance for these species to eat? Not much. And so they stay off it altogether.

Several mammal species have close relatives across the tree line, sort of like the way it used to be in East and West Berlin, only here they are kept apart on ecological, not ideological, grounds. The ranges of the snowshoe hare and arctic hare match up almost as cleanly as pieces of a jigsaw puzzle. So it is with the red squirrel and arctic ground squirrel, the red fox and arctic fox, and the black bear and grizzly bear.

These species do mix around the tree line. Tongues of shrub tundra often extend far south of the forest frontier, especially along high ridges and domed hills. Here, a lemming might just as easily end up between the jaws of a red fox as an arctic fox. I discovered first hand that black bears and grizzlies may also overlap in such terrain. On two separate occasions, while wearing a canoe on Pike's Portage, I met them both.

I take a deep breath, then leap over the side of my esker lookout. I land on the esker's wounded brow, which dips steeply into Humpy Lake and disappears. The tracks of caribou and wolf merge together on the sandy shore. I follow them for a while, then climb back up onto the esker to a spot as flat and wide as a soccer field. An even carpet of crimson bearberry plants is sliced through by several well-worn caribou trails. Their droppings cover the ground like so many hailstones. I select a trail that leads down into a tiny forest of spruce.

This is probably as calm as it gets out here on the tree line — no bugs, no wind, no driving snow. The silence is, as they say, palpable. At least until I hear a distinct thump behind me. The footfall of a grizzly? One of those gnarly old trees finally packed it in as I walked by? My own heartbeat?

A solitary female caribou looks down on me from the soccer field. This trip, they always seem to be looking down on us when we least expect it. Caribou are not renowned for stealth behaviour. To feed themselves there is no need to creep up on lichens and ambush them. Still, there is a twinkle in her eye that makes me feel as if *I* am the hunted.

She looks at me, then over her shoulder, then back at me. She suddenly raises her head and tail, does a quick 90° turn and prances off down the esker. I watch her until she disappears into another clump of trees. I hear shuffling down the trail where the caribou had been. Grizzlies are not known for stealth either. Is *this* one?

My wife and young daughter appear over the crest of the esker. They are carrying buckets. It's time for some serious berry picking.

Tomorrow's Landscape

Introduction

Scientific extrapolations, educated guesses and fervent hopes provide the lenses through which to decipher the unwritten story of the taiga shield's future. Neither fiction nor fact, this story looks at the possible nature and causes of environmental change in a landscape poised on the wilderness fringe of civilization. Learning to care for this landscape and fully participate in its perpetual unfolding is the challenge explored in the last chapter.

The boreal regions are too important to the future of mankind to allow their use to be governed by only one preconceived economic system. The only valid criteria for determining use are ecological ones.

— William Pruitt[118]

21

The Shield Lies Waiting:
Imprints of Future Development

During a sight-seeing tour of the Arctic in 1938, Mitchell Hepburn, then premier of Ontario, had a vision:

> If this gold-rush goes ahead at top speed, the North will soon absorb the unemployed.
> In a few years, there will be great cities on Great Slave Lake.[119]

Although the world was still reeling in the wake of an economic depression and teetering towards the brink of another world war, to Hepburn and many other North watchers, the future looked rosy for this part of Canada. After all, in the space of only three years, a few scattered prospecting camps had blossomed into a bustling industrial town whose workers had already mined, processed and shipped out a million dollars' worth of gold. Hepburn's vision was not too farfetched for his day. But all forecasts are notoriously risky. Over 60 years have passed and the "great cities" have yet to be built.

Though Yellowknife remains the western taiga shield's only city, mining has indeed become its number-one industry. Since the 1930s, scores of gold mines have been built here. Most of them closed up shop once the richest ore was gone. Of the seven operational mines in the Northwest Territories, four of them are planted on the taiga shield. In ten or twenty years these will also close. No doubt new ones will open. Over the next century, they'll be mining new deposits of gold and silver, base metals like lead and zinc and rare "high-tech" metals such as lithium and thorium — and it looks like diamonds too. By July 1993, over 16 million hectares of the Slave and Bear geological provinces had been staked by an unprecedented rush of diamond prospectors.

When present or future mines close down, as all of them do, each will leave a different signature on the shield landscape. Depending on factors such as the nature and depth of

What rich ore bodies lurk in the bowels of the northwestern Canadian Shield? Once discovered, how many of these will warrant the construction of new mines and supporting infrastructure, such as roads and hydro projects? How will the environment fare if this region becomes the mining heartland of Canada? Only time will tell. . . . *(Busse /NWT Archives)*

the ore, the technology used to process it and the way it goes to market, a mine could leave behind a gaping open pit, a small lake or two full of tailings, a gravel airstrip or maybe even an all-weather road or railway line. Until northern air pollution regulations are significantly tightened up, local denudations of the landscape from acid-belching mill stacks could become a spreading problem. Accidental spills or leaky tailings dams may continue to ring occasional alarm bells in the offices of government officials entrusted to protect the exceptional quality of northern waters.

Every landscape offers different resource opportunities, or "gifts," as I like to think of them. Northern shield rock is endowed with some of the most promising mineral prospects on earth. This region could well become the mining heartland of Canada in the twenty-first century. Inevitably, another wave of mines will come. In the process, the landscape will change. Environmental costs will be levied. Measured against the last 60 years of this region's history, I like to believe that these costs will be, as they say, "minimal."

I am convinced that the "bad old days" of northern mining are behind us. No more shall toxic loadings from mines reach such critical levels that the health of entire communities comes into question, as it did in Yellowknife during the mid-1970s. No more shall accidental spills of toxic mine tailings be allowed to continue unchecked and unpenalized, as it did at Discovery Mine on Giauque Lake during the mid-1960s. And no more shall mining companies, big or small, walk away from a mine site without cleaning up or restoring the land, as American Uranium did when the money ran out on Ray Rock Mine in 1959. With improved environmental regulations, advancements in anti-pollution technology and environmentally progressive management regimes established through native land claim settlements, such mining misbehaviour will, I trust, become impossible.

I confess that my confidence in the "system" does waver now and then. During moments of doubt — even fear — about the impacts of future mining on shield country, I am consoled by the fact that spatially most mines are mere specks in the immense wildness of this landscape. Yellowknife geologist John Brophy has translated this notion into numbers. He figures that the average mining operation up here has an "impact area" of 10 square kilometres. Using this number, he estimates that all the current mines in the Northwest Territories have an impact on 0.002 percent, or 1/50,000, of the landscape. His conclusion: "I think the land is big enough to sustain this use."

Brophy's point is that the worst impacts of northern mines are relatively localized. But true to the first law of ecology — everything leaks — all mines inevitably create indirect, cumulative impacts on downstream water and air quality. These impacts are often difficult to measure, let alone control. However, because minerals are non-

Discovery Mine on Giauque Lake, 1953. A tiny island of development from which a horrendous spill of mine tailings sent widespread and long-lasting reverberations downstream. *(Busse /NWT Archives)*

renewable — once they're gone, they're gone — both the direct and indirect impacts of mines can be relatively short lived. That is, of course, assuming that mining companies don't leave behind leaky piles of toxic or radioactive tailings, gaping open pits in the ground or a network of permanent roads and trails slicing through the wilderness. Mines will continue to boom, then bust. Only through far-sighted land-use planning, responsible corporate behaviour and sharp regulatory teeth will the landscape have any hope of recovery once they're gone.

What about this region's major renewable resources, namely, fish, furs and game animals? More intensive use of these resources is inevitable based on population growth alone. For better or worse, the Northwest Territories has the highest annual birthrate in Canada — almost 3 percent, compared to a national average of 1.5 percent. Between 1965 and 1990, shield country's small, traditional communities such as Lutsel'ke (Snowdrift), Rae Lakes and Rae-Edzo tripled in size.

For most people in these communities, wild country food, including fish, caribou and waterfowl, makes up at least 25 percent of their daily fare. Regardless of perpetual ups and downs in the fur market, many still trap marten, mink and beaver in rhythm with the seasons. For food, profit or simply the affirmation of culture, hunting and trapping around these communities will carry on well into the next century, likely in proportion to their growing human numbers.

Superimposed on increased domestic use of fish and wildlife will be a trend towards much intensified commercialization. More fresh meat will be delivered to local store-fronts as well as southern markets hungry for exotic northern specialties. The barren-ground caribou that make an annual traverse through shield country and the deep-water

The traditional hunting and trapping way of life persisted throughout the twentieth century in spite of chronic economic odds against it. Will economic, ecological and cultural priorities for this region favour the persistence of this tradition throughout the next century? Only the great grandchildren of trappers like these will know. From left to right (adults only): Morris Tinkwe, Francis Williak and Philip Nitsiza near Fort Rae, 1949. (*Busse/NWT Archives*)

fish species of the region's great lakes are among the "commodities" slated for increased commercial harvesting.

Taken to extremes — in other words, the limits of biological sustainability — these trends could create problems for some fish and wildlife populations. For instance, commercial fishing of lake trout in Great Slave Lake is already a high-profile resource management issue, one that may well heat up over the next few decades as more people compete for less fish. On the other hand, commercial quotas on caribou are traditionally so small — usually around half of one percent of a particular herd — that, unless jacked up substantially, increased quotas likely will have a negligible impact on their numbers.

The aurora borealis. Tourists come here from around the world to glimpse its splendour. (*Busse/NWT Archives*)

By convention, the more hunting, fishing or trapping pressures a population is subjected to, the more attention it gets from those responsible for monitoring its numbers, researching its biology, protecting its habitat and regulating its take. Presuming that this convention will be upheld into the twenty-first century and beyond, I look upon the long-term health of shield country's fish and wildlife with cautious optimism. I take further solace in the fact that, as the nineteenth century closed, many wildlife species may have been scarcer than they are today due to totally unrestricted harvesting, particularly around the region's forts and trading posts. Back then, there was even talk of the caribou going extinct. Will there be such talk a century from now?

As long as caribou and other natural emblems of the wild northern shield shall last, tourists from all over the world will continue to visit this region. Naturalists, sport hunters, anglers, adventurers, hikers and canoeists — they'll come drawn by travel brochures that speak of an "awesome primeval landscape," countless unspoiled rivers and lakes "teeming with trout" and "incredibly rich wildlife." Among our "enviable tourism resources" are the aurora borealis in winter and the midnight sun in summer, phenomena that some tourists will sprint halfway around the world just to see.

Shield country now hosts more visitors each year than any other region in the Northwest Territories. In 1988, almost 12,000 people vacationed here — up more than 25 percent in less than a decade. The tourism industry hopes to double that number by the turn of the century. At this rate of increase, well over half a million tourists could pay a visit during the year 2050.

That's a lot of tourists. Can the land sustain such swelling numbers of visitors and still keep them coming back? Will shield country go the way of the Himalayas in Nepal, the hill country in Thailand and the remnant jungles in India, where excessive tourism development now threatens the very resources tourists have come to experience? It's time for more cautious optimism.

Almost 200,000 tourists in search of wild country visited the Yukon in 1991. Chances are they found it, since, in spite of tremendous annual visitation, the Yukon's wilderness character remains largely intact. There is no reason to believe that it will be any different here when we begin seeing this kind of tourist traffic in the next century.

As the world rapidly urbanizes, the demand for experiencing nature in as pure a form as possible will multiply. The growing "ecotourism" movement is an early expression of this trend. Getting people out to take a dog-sled ride, watch some caribou, visit a tern colony, find some solitude or catch (then release) a monster trout creates little environmental abuse. Most tourist parties on the land these days are low-key affairs — scattered in distribution, small in numbers and limited by a short season. Future tourist development is unlikely to spoil the shield hinterland so long as it follows this style and builds on local knowledge and sensitivity to the land. Tourism in turn will offer profitable *and* sustainable economic opportunities to the region's communities and unparallelled refreshment to city-weary southerners.

Over the next 100 years, many wild places elsewhere will be either "Disneyized" beyond recognition or wiped off the face of the earth. The tranquil, pristine beauty of the taiga shield will live on. Long-term protection of its wilderness character will be guaranteed by an environmentally enlightened (and controlled) mining industry, a thriving nature-based tourism industry and widespread recognition of the fact that this region is one of the last crown jewels among the planet's dwindling natural treasures. Or so I humbly hope.

22

Global Connections:
The Greenhouse Effect and
Other Influences from Afar

The human population explosion with its accompanying technological explosion is disrupting the orderly development of the world's biosphere in a variety of ways . . . the greenhouse effect appears to be the paramount disruption.[120]

E. C. Pielou, whose life's work has been to study ecological change on this continent since the last ice age, calls the greenhouse effect a "disruption." Some scientists call it a "climatic aberration." Jonathan Weiner, best-selling author of *The Next Hundred Years*, calls it a "global catastrophe."[121] However they look at it, authorities seem to agree on two points. First, warming of the earth's atmosphere by gases released from industrial development is inevitable. In fact, there is growing evidence that this warming has already begun. And second, it will influence different regions in different ways.

Working much like a clear pane of glass, our atmosphere allows high-energy shortwave radiation from the sun to reach the earth's surface and heat it up. If this energy were not somehow reradiated back into outer space, the earth would slowly but surely be converted into a fireball. Fortunately, an equivalent amount of energy is given off by the earth as longwave infrared radiation. Before it gets a chance to escape, certain atmospheric gases, particularly carbon dioxide, absorb some of this outgoing energy and bounce it back down to earth, only to reradiate it. In the grand history of the earth's climate, this process is all very natural.

What is unusual, and somewhat alarming, is the fact that over the short span of two centuries our species alone has managed to raise the level of atmospheric carbon dioxide by 30 percent through the burning of carbon-based fuels and massive destruction of the earth's forests, which once served as giant reservoirs of stored carbon. More carbon injected into the atmosphere means more absorption of outgoing infrared energy. More energy absorption means things heat up here on earth, just like a greenhouse should.

The fastest, most pronounced warming on earth will likely happen in northern latitudes. Ironically, snow and ice will have a lot to do with this trend. Because they are highly reflective, a huge proportion of solar radiation bounces back into space before it can warm the earth. But as the insulating blanket of carbon dioxide builds overhead, more and more of this reflected radiation will tend to be trapped by the atmosphere, an effect greatly enhanced near the earth's surface by prolonged temperature inversions (warm air below, cold up top), which often lock in for weeks beneath the arctic air mass.[122]

On the taiga shield, some of the likely effects of this warming will be positive, at least from a human perspective. Who, for instance, could complain about the prospect of warmer, shorter winters? Sometime between the year 2030 and 2050, when atmospheric carbon dioxide is expected to double, average winter temperatures in this region could climb as much as 10°C. The change in summer temperatures will be less dramatic — in the order of 2 to 5°C — but probably enough to lengthen the growing season by at least six weeks. Large areas of permafrost will likely disappear, reducing future construction and engineering headaches created by frost heaving. Water bodies could remain ice free for 30 to 40 days longer, which would extend opportunities for water-based recreation and transportation. Residents of Yellowknife will likely enjoy the kind of climate that Edmonton, 1,400 kilometres south, gets today. If so inclined, they could grow wheat, corn and beefsteak tomatoes in their unprotected backyard gardens.

It looks inviting. But, Barrie Maxwell, a meteorologist from Environment Canada, warns that not all the effects of warming will necessarily be beneficial, neither for us nor for the animals and plants that live here.

> We should not allow the general belief that future climate change is likely to warm up the Arctic delude us into thinking that everything will suddenly be easier to deal with in that region. There are advantages and disadvantages to be weighed. . . .[123]

At best, the greenhouse effect will be a mixed blessing for this region. At worst, it will result in the unravelling of an entire ecosystem.

Some predictive models — and there are many — suggest that, of all the world's major ecosystems, taiga forests will be the most affected by greenhouse warming. Many characteristics of these forests, including species composition, biological productivity, soil conditions, fire, insects and disease, are all inextricably linked to climate. When temperatures begin to rise, these forests may change in several dramatic ways.

The most common natural disturbance to the taiga is fire. Some ecologists predict that fire's ancient balancing act between destroying the forest and rejuvenating it will be upset by greenhouse warming. In the next century, the balance may tip towards destruction.

If the climate changes as predicted, wildfires will likely become more frequent and intense due to drier forests and soils plus a longer burning season. Some predict that changes in global circulation patterns caused by relatively sudden warming will bring more unstable, stormier weather. For summer in this region that means increased lightning, which starts most forest fires. It also means more wind, which spreads them around.

Massive infestations of spruce budworms, sawflies, bark beetles and other insect pests are virtually unknown in northern coniferous forests — at least up till now. Outbreaks of pests will likely spread north as temperatures rise and the season for their attack gains time at both ends. Trees stressed by relatively rapid shifts in temperature and parched soils may be particularly susceptible. A recent series of warm summers may be partly responsible for a spreading infestation of spruce budworms in the Slave River Valley. Is this the beginning of a long-term trend?[124]

However the forests fall — by fire, insects or disease — in some situations they may not recover. A host of ecological changes brought by warming may create conditions to which the forest cannot adapt: the drying and decomposition of peaty soils, the degradation of permafrost, which on slopes could result in much soil slumping and mass erosion, and the exposure of extensive areas of mineral soil through more frequent and intense ground fires.

Vegetation better adapted to these altered conditions may move in and assert itself. Southern parts of this region may eventually become dominated not by spruce and pine forests but by aspen parklands or dry grassy steppes. On many high rocky ridges, vegetation may not return at all due to the loss of soils, either burned off completely or washed away during more violent spring runoffs. In this way, more shield rock may gradually become exposed to the sunlight and remain bare, dry and hot.

The specific effects of greenhouse warming on the forests of this region will vary from site to site depending on a myriad of ecological factors, such as drainage, soil depth, slope, microclimate and seed sources. One general effect that ecologists seem to agree on is that the tree line will once again head north, as it has several times since the last ice age. The northern fringe of coniferous trees could advance out onto the tundra by 200 to 300 kilometres. The southern fringe, on the other hand, could move north by as much as 950 kilometres, which would result in a significant net loss of taiga forest.[125]

Some estimates predict a gradual nibbling away of the taiga from the south as ever-increasing thresholds of greenhouse warming are crossed. Consensus is that in many areas it will likely become patchy or disappear altogether. I rub my eyes in disbelief when I read, in reputable scientific papers, that a 99 percent loss of taiga forests is not improbable.[126]

Gross changes to vegetation caused by the greenhouse effect will not happen overnight. Even if the warming comes swiftly, there may be a time lag of 100 years or more before a new ecological order is established. For instance, northward migration of the tree line will depend on more than just warmer, longer summers. Until the right changes occur in tundra soils, trees will not be able to advance onto them and take root. Similarly, coniferous forests in the southern part of this region will probably be able to hang on for a generation or so even after they become unsuited to the changing climate.

What does all this mean for the region's wildlife? One of Canada's preeminent taiga ecologists, William Pruitt, has pondered such questions. The answers, he concludes, must be considered as "third-order extrapolations which are, if you will pardon the pun, quite hairy."[127] Still, in the name of scientific speculation, he puts forward some interesting predictions. According to Pruitt, the greenhouse effect may be good news for garter snakes, which may be able to expand their range into more northerly parts of the taiga shield. As for frogs and other amphibians, range expansions seem less likely since the advantage of warmer temperatures may be offset by the disadvantage of drier habitats. Birds of open habitats such as flycatchers and sparrows will probably be favoured over species preferring thick conifer forests. Pruitt feels that woodpeckers and other bird species benefitting from dead snags "will probably undergo population explosions as long as forests remain to burn and die."[128]

He predicts a decline in mammals closely associated with coniferous forests such as marten, red squirrel and lynx. Snowshoe hare will probably become more common as the proportion of deciduous growth increases. Pruitt adds that "it would be foolish to speculate about whether the hare cycle will continue or not."[129] The numbers and range of Canada's most common mammal, the northern red-backed vole, will probably decline. Drying in the southern part of this region will probably favour an increase in small mammals adapted to steppelike conditions, such as the meadow vole.

No doubt barren-ground caribou will remain as unpredictable as ever. Pruitt's hunch is that they may suffer from a major reduction of the southern portion of their winter range. For the western taiga shield region, this could mean a shrinkage in the so-called "caribou range" area south and east of Great Slave Lake, which now is used extensively by the Beverly caribou herd, currently numbering around 200,000 animals. Although winters will be shorter, snowfall will increase — perhaps by up to 22 percent — which means that caribou will have to work much harder to locate food. Not only will the snow be deeper, but more frequent invasions of warm, moist air from the south may result in crustier, ice-layered snow, which could make travelling more difficult for caribou or lower their odds for escaping the jaws of a wolf.

A greater incidence of ice-layered snow also could create problems for small mammals living under it. Like a lid on a bottle of soda pop, these layers tend to prevent the escape of carbon dioxide, which builds up beneath the snow in late winter from microbial activity in the soil. In high concentrations, this gas can poison small mammals or force them to the surface, where they risk hypothermia and increased chances of predation. Theoretically, frequent and widespread icings of the snow cover could lower the overall abundance of small mammals and send negative ripples through all the animal food chains depending on them.

And what's the prognosis for fish? Not good, according to a team of scientists led by David Schindler, a biologist with the University of Alberta in Edmonton. For almost 20 years, his team has studied the effects of climatic warming on shield lakes in northern Ontario. Over this time, they measured changes in just about every ecological variable they could think of, such as air temperature, lake temperature, precipitation, ice-free season, wind speed, spring runoff, phytoplankton and so on. The results of their labour provide "a preview of how climatic change may affect boreal lakes and catchments in the next century."[130] One of their most disturbing discoveries was that a number of cold-adapted species, including lake trout, may be "locally extirpated" — they may disappear — from many smaller lakes as temperatures rise.

The likely problem of declining lake trout in shield lakes provides a good illustration of the complex ecological interactions our species has unwittingly set into motion because of the greenhouse effect. During the course of Schindler's lake study from 1969 to 1987, many of the changes expected from climatic warming were recorded: a rise in air temperature of about 2°C, an increase in the frequency and intensity of forest fires, stronger winds, a longer ice-free season and an earlier disappearance of snow. A combination of these factors contributed to increased warming and thermal mixing of lake waters. The scientists predict that in some shield lakes the cold, oxygen-rich bottom layers, which lake trout depend on in summer, may shrink significantly or disappear altogether.

Once the trout and other cold-water species are gone from a lake, there is no guarantee that other warm-water species, such as northern pike or walleye, will simply move in and take over the empty habitat. On this note, the scientists are not optimistic. "Even though the extirpated fauna might be replaced by warm-water assemblages, it is by no means certain that fisheries of comparable value or ecosystems of comparable diversity would be reestablished quickly."[131]

The greenhouse effect will likely have other impacts on shield lakes, caused not only by higher temperatures but also by increased pollution loads dumped from the sky and the land. More frequent winter storms from the south could bring with them polluted air masses originating in southern Canada and the United States. Meanwhile, increased disturbances to the north could stir up largely Eurasian pollutants suspended in the atmosphere as "arctic haze." The combined effect will likely be more acid precipitation, falling mostly as sour snow.

Shield lakes are notoriously sensitive to acid precipitation. The main reason is that much of the granitic bedrock around and under them is already acidic. Unlike more alkaline sedimentary rocks in the Mackenzie Valley, shield rock generally has no buffering capacity to neutralize an additional load of acid. Most of this loading will likely happen suddenly during spring runoff, which is predicted to be quite a dramatic event given the increased snow volumes and warmer spring temperatures. Ecologically, this "acid shock" could not come at a worse time.

With the arrival of spring, the reproductive cycles of most aquatic species enter full swing. It is also the time when parts of lakes or rivers may become as much as 100 times more acidic in just a few days or weeks. Such levels in remote shield lakes have been recorded during snowmelts in northern Ontario. It turns out that lake trout, once again, are particularly sensitive to this phenomenon and they are usually among the first species to go. With the disappearance of more and more species — fish, invertebrates, plants and plankton — scores of lakes have already been pronounced dead.

Is this a foreshadowing of what's in store for some lakes on the taiga shield? If, as seems likely, our winters are destined to become progressively more soured by acid fallout from afar, then the answer is clear — as crystal clear as the waters of a dead lake.

The land will also suffer if acid levels rise significantly. The brunt of the impact will be taken by soils and lichens. Soils are vulnerable for the same reason lakes are — most of them rest on acid rock. Most taiga soils are quite acidic as it is. Any increase could harm soil microorganisms, lower soil fertility and interfere with the uptake of soil nutrients by plants. This could add up to great stress on the entire ecosystem.

As for lichens, few forms of life are more sensitive to acid fallout. Because they lack roots, lichens must pump nutrients directly from the air. Though well adapted for this seemingly miraculous task, lichens have no ability to filter out airborne acids such as sulphur dioxide and nitrous oxides. As well, lichens lack a protective cuticle layer — standard equipment on all leaves and needles — so that after an acid bath in rain or meltwater, they become coated in pollutants, which they have no option but to absorb. Once inside, acids can strangle the lichen's life-support system by slowing down both photosynthesis and respiration.[132]

Lichens are so sensitive to acid fallout that they have been used as bioindicators for mapping pollution zones around urban and industrial sites. Around Yellowknife, such a study measured the impacts of acid emissions from gold smelting and discovered symptoms of stress in lichens and other vegetation as far away as 25 kilometres downwind from the main stack.[133] Closer to town, many species of lichens common to this region were totally absent. Others were stunted or in the final stages of disintegration. The surviving lichens, those labelled "toxitolerant," were few in number compared to the rich variety of species found throughout most of the taiga shield.

In the next century, will acid fallout from both local and remote sources result in widespread impoverishment of shield country's lichens? If so, more than the lichens will suffer. Lichens feed caribou, build soils, buffer water erosion and nurture pioneering plants. As elfin beauty marks on the shield, they also nurture our souls. So, in a sense, we all may suffer with their decline. Faced with such a prospect, I sometimes run low on optimism.

Spring melt could become hazardous to the health of many aquatic organisms if long-range airborne pollutants increase, thus raising the risk of acid shock. *(John Poirier)*

Darker still, yet less tenable, are my concerns about potential impacts of another danger flung north by the industrialized world: the depletion of atmospheric ozone, which shields living things from the sun's hazardous ultraviolet rays. Eating away at the ozone layer are a number of synthetic chemicals, the most destructive of which are CFCs, chlorofluorocarbons, which cool our refrigerators and air conditioners, insulate our homes and offices and cushion our chairs and car seats.

In 1985, the infamous ozone hole over Antarctica was discovered. In 1986 and 1989, similar holes showed up over northern Europe. At that time, Environment Canada issued a public fact sheet that downplayed, rather whimsically, the possible hazards of ozone depletion in northern areas:

> Scientists feel there is no increased risk to people living in the Arctic. The ozone layer over the far north is naturally very thick, and to date, only a small amount has been lost. . . . Depletions in Arctic zones have been observed only in spring, during February and March — a time of year when northerners would not be sunbathing.[134]

In February 1992, startling new data on pollutant levels in the upper atmosphere led to the prediction that a huge ozone hole, comparable to that over Antarctica, might soon open over all of northern North America and much of Eurasia.[135] By March the same year, Environment Canada officially changed its tune by issuing the first of its daily "UV watches" to help keep the public regularly posted on increased radiation hazards contributed to by ozone depletion.

Apparently my cherished collection of baseball caps is now out of vogue. If I insist on wearing them in future, the exposed tops of my ears run the risk of getting skin cancer. As the ozone layer begins to thin over our heads, wide-brimmed hats — not to mention long-sleeved shirts and plenty of sun screen lotion — will, by necessity, become standard northern apparel for those who dare spend a lot of time outdoors in the spring and summer. More people, from toddlers on up, will wear dark, UV-filtering sunglasses to reduce the likelihood of going blind from cataracts caused by increased ultraviolet radiation.

Most of these precautions are already being taken by sheep herders in southern Chile, which lies on the outer fringes of the well-established Antarctic ozone hole. Unfortunately no protective gear has been designed for their sheep, some of which have already gone blind from an overdose of UV rays.

How quickly this crisis will reach similar, if not greater, proportions in the Northern Hemisphere is unknown. "Maybe this spring, maybe next," is the best prediction scientists can give us. When it comes, what will be the specific impacts of ozone depletion on the taiga shield? How long before the first barren-ground caribou or red-throated loon goes blind, like the sheep in Chile? How might increased UV radiation influence plant growth, particularly those thin-skinned lichens? In the region's many lakes and rivers, how might it damage sensitive organisms, such as phytoplankton, which form the base of many aquatic food chains? And might we humans — sunbathers or not — notice a marked increase in skin cancer and our susceptibility to infectious diseases? No one can say.

Some scientists feel that northern Canada will be among the areas hardest hit by ozone depletion. Speaking to a *News North* reporter in Yellowknife, Dr. Allan Clark, an ozone specialist at the University of Calgary, predicted that "When this problem gets worse, the hole will get to you people first."[136] Others claim that we may in fact be better off than affected areas to the south since UV rays become more diffuse, hence weaker, with increasing latitude. Still in its early stages, this debate goes on, overshadowed by a thick cloud of uncertainty.

Attempts to accurately predict impacts on the taiga shield and other northern regions are thwarted by the fact that scientists have no past experience to build on. Nothing like this has ever come north. That it will come seems certain. Millions of tons of ozone-destroying chemicals are still being released each year. According to Clark, "Even if we stop using the damned stuff right now, the atmosphere still will have to deal with it for the next 40 years."[137]

The greenhouse effect, acid fallout and ozone depletion are pervasive threats that may tar all of shield country with the brush of a heavily urbanized civilization thousands of kilometres away. I hope that many of my concerns about the impacts they may bring to this landscape will be proven groundless as time rolls on. Which is more naive: to worry about future ecological tragedies imposed from afar on this wild and wonderful land, or to believe that some day, very soon, our species may wise up in time to prevent them?

23

Living Locally and Lightly:
A Bioregional Perspective

We were all native to some place at some time in our ancestral past, and it is our natural right to become native to some place again.[138]

Eighteen nights and only nine suppers' worth of protein — during this canoe trip, the main dish for half of our evening meals was to be supplied by the land. The concept made both practical and philosophical sense as we stuffed our seemingly bottomless packs in Yellowknife. Every item we elected to leave behind, from Polish sausage to lentils, meant less back-breaking work along the many portages ahead. Besides, *having* to catch our supper would bring us closer to the place through which we were journeying: across the tree line from the East Arm of Great Slave Lake to Artillery Lake on the barren lands.

After my hundredth cast into the mouth of the Lockhart River, the idea of living off the land, even partly, had lost most of its charm. In my stomach was a large hole into which a fish would neatly slide. In my head were mildly disturbing visions of tonight's dinner plates: East Indian white rice and dehydrated California carrots sitting utterly alone. Apparently taiga shield lake trout was not on the menu.

I had tried the full gamut of my proven trout teasers and aimed them all at surefire riffles and pools where the fish should have been. The fruits of my labour were two suspected nibbles and four snagged lures lost forever from my prized collection. As a last resort, I tried a goggle-eyed copper and silver spoon that, in Ontario, had caught many a black bass at the hand of my grandfather but up north had caught nothing but weeds. Judging by the growing hole in my stomach, I decided to allow myself only three desperate casts. On the third and final cast, fate smiled. Within minutes I had three pearly trout on my stringer.

My opportunities to experience a deep, visceral dependency on this land are rare. As a city-slicker from Yellowknife, I belong to a transplanted society that excels in exploiting things — food, materials, energy — from elsewhere and elsewhen. It thrives on an economy that inherently ignores our dependency on the natural world. It is based on a homogenized consumer culture rooted nowhere, since it is found everywhere.

Still, I do my best to bring resources at hand into my daily life. I put local wild food on my table whenever I can: cranberries picked from a spruce-feathermoss forest, caribou taken during a fall hunt at the tree line, fish netted from Great Slave Lake and syrup refined over a campfire in a friend's secret birch grove. Though mere embellishments compared to the truckloads of staples we consume from the South, this food is held most dear, since it is freely given by this land.

My dwelling too is, in small but significant ways, a product of this land. Each stud and joist of my house is imprinted with a stylized silhouette of a caribou, telling me that I am surrounded on all sides by home-grown spruce. Below me is a metre-thick foundation of crushed volcanic rock, so-called "mine muck," brought to the light from the bowels of the Canadian Shield. In winter, the brief, low-angled subarctic sunlight adds a small but tangible measure of passive solar warmth to my home. And in summer, native trees planted in my humble eco-garden cast welcome shade on my windows.

More than bringing me closer to some abstract notion of "nature" or bolstering my sense of becoming more "self-sufficient," these modest links to the land help establish my kinship with the specific place I have chosen to live — the taiga shield.

The concept of "living in place" is not new. The original inhabitants of this land as we know it, the Athapaskans, spent hundreds of generations learning how to live here: learning the pattern of caribou migrations, the life cycles of whitefish and lake trout, the blessings and perils of forest fires, the best water and overland routes to the barren lands, the many properties of subarctic snow, and so much more. Their profound knowledge of the land was reflected in their tools, their shelters, their seasonal movements and, no doubt, their songs. Their community included the trees, birds, animals and rocks of the region. This land was their life, as it still is today for many of their Dogrib and Chipewyan descendants. Although they didn't know it, the original people of this land were, in a sense, model bioregionalists.[139]

To practise bioregionalism, one must become a true "dweller in the land," not necessarily to the Athapaskan extreme — those days are over — but to the extent that we can ground our experience, our knowledge and our wealth in the place that we live. According to Kirkpatrick Sale, an early pioneer of the bioregional movement, learning again how to live in place is one of the most important tasks we as a species have before us.

> The crucial and perhaps all-encompassing task is to understand the place, the immediate, specific place, where we live. . . . We must somehow live as close to it as possible, be in touch with its particular soils, its waters, its winds. We must learn its ways, capacities, its limits. We must make its rhythms our patterns, its laws our guide, its fruits our bounty.[140]

Bioregionalism calls for sustainable ways of life rooted to the unique opportunities, pleasures and constraints presented by a particular natural region. In pieces, it means *bio* – life, *regional* – a certain place, and *ism* – the science and art of living as part of a *bioregion.* Simply stated, it means becoming fully and responsibly alive in your natural home.

Bioregionalism means more than tinkering with fifty simple ways to save the planet. While the planet does need our attention — especially when many of the threats to this region's health are planetary in scale — the idea is to think globally and act locally. In theory, the ecological wisdom and connectedness that one gains from a bioregional way of life inevitably ripples out into the world, whether over a backyard fence, into the local council chambers or across international borders. I like this theory.

The bioregional path begins where your feet touch the ground. Getting to know the boundaries of your home bioregion is one of the first steps. If it happens to be the taiga shield, this step is easy. By North American standards, this region is remarkably well defined by the tree line to the north and east, the edge of the Canadian Shield to the west and the closed boreal forest to south. Not surprisingly, this boundary corresponds almost perfectly with the areas called home by the Dogrib and Chipewyan natives.

The next step — a long one — is to learn about your bioregion's unique creation story and discover how it ticks today. Whether you live on the taiga shield, are just visiting it or simply wonder about it, this book was meant to help bring you farther along in this direction — but I see now that there is so much farther to go!

While becoming more conscious of the life and times of my neighbours — the rocks, plants, animals and all else that collectively make this region unique in all the world — I have put down psychic roots that penetrate deep into the Canadian Shield. My sense of place and belonging springs from these roots. So does my appreciation of time and mystery.

Perhaps eventually I will learn what it really means to be a dweller in this land. Ask me in twenty years. I am not native to this place but am gradually becoming so. In the meantime, I will strive to nurture a way of life based not so much on living off the land, but living in the land and it in me.

(John Poirier)

Notes

[1] Gandhi, 1982:52.

Introduction

[2] Lopez, 1987:204.
[3] The taiga shield is one of 15 terrestrial ecozones identified by Environment Canada, 1986. See Pruitt, 1978, for an excellent overview of taiga forest ecology from a circumpolar perspective.
[4] The western portion of the taiga shield ecozone includes the central continental Northwest Territories, plus portions of northern Alberta, Saskatchewan and Manitoba.

An Eagle's View

[5] Henry, 1988:9.

The Making of a Landscape

[6] Moon, 1970:8.
[7] Henderson and Jolliffe, 1939:316.
[8] See Morell, 1990:26, for a general overview of this discovery. Bowring *et al.*, 1989, provide a more technical perspective.
[9] Well-preserved evidence from the Earth's moon, Mars and Mercury suggest that all the inner planets of our solar system suffered several periods of bombardment by smaller "planetesimals" from the time they formed until about 3.8 billion years ago (Smith, 1981:250).
[10] Padgham, 1987:1.
[11] Padgham, 1991:7.
[12] See Padgham, 1990, for an excellent overview of current theories on the creation, movement and transformation of the Slave geological province.
[13] Price, 1967:141–145, recounts the wild stories of old-time prospectors and greenhorn amateurs stampeding into the bush in response to the 1938 "Treasure Island" gold discovery on Thompson Lake 50 kilometres northeast of Yellowknife.

[14] Price, 1967:237.

[15] See Gore, 1989:674–675, for a rare look at *living* colonies of these organisms.

[16] Levenson, 1989:12. See chapter entitled "In the Beginning" for further discussion of the evolution of the earth's atmosphere.

[17] This story is neatly summarized in Redfern, 1983:26–28.

[18] Quoted in Roberts and Amidon, 1991:146.

[19] See Smith, 1981: chapter 18, for a fascinating overview of the climatic history of the earth.

[20] Janzen, 1990:6.

[21] See Thurston, 1986, for further details on northern life and climate during this period.

[22] See Gore, 1989, for a clear synopsis of the dinosaurs' decline.

[23] Stonehouse, 1971:44–45. See also Pielou, 1991:147–163.

[24] Pielou, 1991:251.

[25] Pielou, 1991:265.

[26] See Craig, 1965, for the original story of the rise and fall of Glacial Lake McConnell.

[27] See Vanderburgh and Smith, 1988, for more details on this transformation of Great Slave Lake.

[28] Bird, 1972:19–20, describes how the erosive action of rivers had a major influence in shaping the Canadian Shield landscape.

[29] Ritchie and Hare, 1980, describe many presumed changes in the tree line during this period.

[30] Pielou, 1991:156–157.

[31] See Garrett, 1988, for an overview of the peopling of North America.

[32] See Rogers and Smith, 1981, for more details on the Shield Archaic culture and movements. Noble, 1981:97–98, describes the Acasta artifacts and distribution of sites.

[33] Government of the Northwest Territories, 1976:1.

[34] Wright, 1981:86–87, proposes a number of "inseparably related physical and cultural phenomena" to explain this archaeological "homogeneity" of shield prehistory.

[35] Yearly cycle derived primarily from Rogers and Smith, 1981.

[36] Moon, 1970:23.

[37] Bromley, 1986, provides an excellent summary of the development of the Northwest Territories' fur trading economy.

[38] Rich, 1968:371–372. Parker (1971:17–18) provides another account of lean times at Fort Chipewyan and describes its absolute dependency on fish resources during the winter.

[39] McFarlane, 1880:808–809.

[40] Janzen, 1990:23.

[41] Cited in Janzen, 1990:18.

[42] Franklin, 1824:221–222.

[43] Camsell, 1954:189.

[44] Bury, 1917, cited in Janzen, 1990:62.

[45] Cited in Graves, 1988:16.

[46] Lower, 1936, cited in Janzen, 1990:43.

[47] Russell, 1898:88.

[48] Cited in Fumoleau, 1973:131.

[49] Cited in Janzen, 1990:44.

[50] Kitto, 1920:88. See Bergerud 1974 for a well reasoned discussion of possible factors contributing to caribou declines.

[51] Barr, 1991:97.

[52] Cited in Fumoleau, 1973:246.

[53] Hewitt, 1916:32–40.

[54] Graves, 1988:61.

[55] Cited in Fumoleau, 1973:238.

[56] Henderson and Jolliffe, 1939:314.

[57] MacDonald, 1942, cited in Janzen, 1990:47.

[58] Laytha, 1939:145.

[59] Laytha, 1939:346, 350.

[60] Laytha, 1939:256.

[61] Laytha, 1939:122.

[62] Laytha, 1939:273–274.

[63] Gould, 1984:115.

[64] Cited in Laytha, 1939:224.

[65] Laytha, 1939:70.

[66] Price, 1967:196.

[67] Finnie, 1948:133–134.

[68] Price, 1967:206.

[69] Kelsall, 1968:283.

[70] Cited in Janzen, 1990:102.

[71] Potts, 1988:1.

[72] Debates of the thirty-first session of the Government of Northwest Territories Legislative Assembly, cited by Graves, 1988:164.

[73] Barr, 1991:100. In 1990, the community approached the territorial government to grant it a muskox quota in view of the recent expansion of the muskox's range into traditional hunting areas east of Artillery Lake.

Today's Landscape — Part 1

[74] Pruitt, 1983:4.

[75] Atmospheric Environment Service, 1986:2.

[76] Atmospheric Environment Service, 1993.

[77] See LaChapelle's *Field Guide to Snow Crystals*, 1992, for a beautifully illustrated description of different kinds of snow crystals and how they metamorphose over time.

[78] Refer to the climate data in Environment Canada, 1990, to see how Yellowknife measures up to other communities across Canada.

[79] See Environment Canada's *Marine Guide to Local Conditions and Forecasts*, 1991, for more information about winds and weather patterns on the region's "great lakes."

[80] In *Canada Moves North,* Richard Finnie (1948:146–168) recounts such stories.

[81] See Tarnocai, 1978, for an excellent review of soil types found in northern Canada.

[82] Brown, 1970:129.

[83] Adapted from Bird, 1972:153.

[84] Quoted in Walker, 1984:12–13.

[85] See Thomson *et al.*, 1969, for a detailed review of lichens in this region.

[86] Larsen, 1980:41.

[87] Lauriault, 1989:103.

[88] Quoted in Walker, 1984:39.

[89] Quoted in Walker, 1984:62.

[90] Quoted in Walker, 1984:121.

[91] Government of the Northwest Territories, 1976:5.

[92] Moon, 1970:128.

[93] It is interesting to note that true hibernators are common to the south (e.g., chipmunks, groundhogs, meadow jumping mice) and north (e.g., arctic ground squirrel) of this region. Why are there none here? No one knows.

[94] Northern News Service, 1991:9.

[95] Quoted in Walker, 1984:16.

[96] Moon, 1970:22.

[97] The description of these fires was derived from Indian and Northern Affairs Canada (1973) forest fire reports archived in the Government of the Northwest Territories' Forest Fire Centre in Fort Smith.

[98] Scotter, 1977:12. See this reference for further discussion of the relationship between forest fires and caribou.

[99] Government of the Northwest Territories, 1991.

Today's Landscape — Part 2

[100] Government of the Northwest Territories, 1993a:22.

[101] Bell, 1902:17.

[102] Plant list derived from Hale, 1979; Ireland *et al.*, 1980; Porsild and Cody, 1980; Thieret, 1964; Scotter, 1966; and direct field observations.

[103] Bird list derived from Bastedo, 1986; Erskine, 1977; Godfrey, 1986; McElroy, 1974; and direct field observations.

[104] Plant list derived from Hale, 1979; Ireland *et al.*, 1980; Porsild and Cody, 1980; Thieret, 1964; Scotter, 1966; and direct field observations.

[105] Bird list derived from Bastedo, 1986; Erskine, 1977; Godfrey, 1986; McElroy, 1974; and direct field observations.

[106] Plant list derived from Hale, 1979; Ireland *et al.*, 1980; Porsild and Cody, 1980; Thieret, 1964; Scotter, 1966; and direct field observations.

[107] Bird list derived from Bastedo, 1986; Erskine, 1977; Godfrey, 1986; McElroy, 1974; and direct field observations.

[108] Plant list derived from Hale, 1979; Ireland *et al.*, 1980; Porsild and Cody, 1980; Thieret, 1964; Scotter, 1966; and direct field observations.

[109] Bird list derived from Bastedo, 1986; Erskine, 1977; Godfrey, 1986; McElroy, 1974; and direct field observations.

[110] Cited by Freden, 1990:71.

[111] Government of the Northwest Territories, 1985:1.

[112] Cited by Raffan and Simmons, 1980:2.

[113] Brophy, 1987:11.

[114] Plant list derived from Hale, 1979; Ireland *et al.*, 1980; Porsild and Cody, 1980; Thieret, 1964; Scotter, 1966; and direct field observations.

[115] Bird list derived from Bastedo, 1986; Erskine, 1977; Godfrey, 1986; McElroy, 1974; and direct field observations.

[116] Cited in Freden, 1990:71.

[117] Except perhaps in exceptionally warm years.

Tomorrow's Landscape

[118] Pruitt, 1989:609.

[119] Cited in Laytha, 1939:233.

[120] Pielou, 1991:311.

[121] Weiner, 1990:213.

[122] See Canadian Arctic Resources Committee, 1987, for further discussion of the potential impacts of global warming on northern Canada.

[123] Canadian Arctic Resources Committee, 1987:6.

[124] A report from the summer of 1991 (The Press Independent, 1991:5) indicates that in locations as far north as Fort Norman spruce budworm infestations are "getting more intense and some of the older trees are starting to die off."

[125] Pruitt, 1989:607.

[126] Pruitt, 1991:6.

[127] Pruitt, 1991:10.

[128] Pruitt, 1991:11.

[129] Pruitt, 1991:12.

[130] Schindler *et al.*, 1990:967.

[131] Schindler *et al.*, 1990:968.

[132] See Lechowicz, 1982, for a more detailed analysis of how lichens respond to acid precipitation.

[133] Hocking *et al.*, 1978:137. See also Government of the Northwest Territories, 1993b, for a more recent discussion of the severity of SO_2 emissions from gold smelting and their impacts on surrounding vegetation.

[134] Environment Canada, 1989:2.

[135] Lemonick, 1992:40.

[136] Northern News Service, 1992:1.

[137] Northern News Service, 1992:2.

[138] Andruss *et al.*, 1990:33. This is an excellent reference on the philosophy and practice of bioregionalism.

[139] Davis, 1993, eloquently describes other examples of how various aboriginal peoples continue the tradition of "living in place."

[140] Sale, 1983:10.

Glossary

Abrasion – A type of glacial erosion in which the rock is worn down, bit by bit, by moving ice or objects embedded in the ice.

Active layer – The top layer of ground subject to annual thawing and freezing in areas underlain by permafrost.

Air mass – A large section of the lower atmosphere with horizontal uniformity of temperature and moisture conditions. Throughout the annual cycle, air masses move in generally predictable ways in response to changing energy inputs from the sun. A rapid transition or "collision" of air masses usually results in stormy weather.

Archean eon – The period spanning from the time the earth's crust first solidified to approximately 2.5 billion years ago — over half the lifetime of this planet. The Archean and *Proterozoic* eons together make up the *Precambrian* era.

Arctic air mass – A large climatic region positioned over arctic North America that is dominated by high-pressure weather systems that are typically very cold, dry and stable. (See *air mass.*)

Athapaskans – Linguistic group of North American aboriginal people with cultural roots extending from the Dene of subarctic Canada to the Navajo of the mid-western United States.

Beringia – A wide, natural causeway or "land bridge" that joined northwestern North America with northeastern Asia. It existed throughout most of the *Mesozoic* and *Cenozoic* eras, except for a break from around 20 million to 5 million years ago. During the *Pleistocene epoch*, this connection opened and closed frequently as sea levels rose and fell in response to the amount of water locked up in glaciers. The last time it closed was around 12,000 years ago.

Biome – A large, easily recognizable region of generally similar climate and vegetation, for example, tundra, boreal forest, desert and tropical rain forest.

Bioregion – A specific natural region characterized by a particular combination of plants, animals, climate, physiography, geology, etc. Ecological and traditional land use patterns often overlap in bioregions hence they represent both geographical and cultural terrains. A subset of a *biome.*

Bog – Bogs are poorly drained areas covered by mats of moss, the most common species of which is *Sphagnum.* Because of high acidity and low levels of oxygen, biological decomposition in bogs is very slow, creating thick underlying layers of *peat.* As peat piles up over many years, bogs often take on a raised or plateaulike appearance. Low shrubs, such as Labrador tea, plus scattered black spruce and willows are the most common plants on northern bogs.

Brunisols – One of nine major soil orders recognized in the Canadian System of Soil Classification. Brunisols are characterized by loosely defined horizon layers that include a distinctive brownish "B" horizon. The most common forms of this soil type in the taiga shield region — dystric brunisols — are affected by *cryoturbation,* though usually to a lesser extent than are *cryosols.*

Cenozoic era – The geological unit of time referring to the era of "recent life." Spanning from 65 million years ago to the present, it is divided into two main periods, the Tertiary, which lasted 63 million years, and the Quaternary, which began 2 million years ago with the *Pleistocene epoch.*

Chatter marks – Small crescent-shaped, steeply inclined fractures created in brittle bedrock by friction from an overriding glacier. Also called "friction cracks."

Continental air mass – A large climatic region positioned over central North America that is generally dominated by low-pressure weather systems. (See *air mass.*)

Continuous permafrost – *Permafrost* occurring everywhere beneath the land surface regardless of vegetation, soil or rock cover. The zone of continuous permafrost begins, more or less, at the tree line and extends northward.

Crown fire – A forest fire that advances through the main upper canopy of tree branches. It may burn in conjunction with a *surface fire* below or it may burn independently, running through the tops of trees without the support of a surface fire.

Cryosols – One of nine major soil orders recognized in the Canadian System of Soil Classification. Cryosols are characterized by a high ice content and marked evidence of *cryoturbation.* They are always associated with *permafrost* within 1 metre of the soil surface (or 2 metres if the soil is strongly cryoturbated). Their high ice content makes them extremely sensitive to surface disturbance. The dominant kind of this soil type in the taiga shield is organic cryosols. In Canada, cryosols cover 3,672,080 square kilometres, or about 40 percent of the country.

Cryoturbation – The regular freezing and thawing action that churns up northern soils and creates patterned ground features such as *mud boils, hummocks* and polygons.

Discontinuous permafrost – Patchy distribution of *permafrost* that depends primarily on the insulating characteristics of vegetation cover and the intensity of solar radiation.

Drunken forest – Forest cover over permafrost where trees lean in random directions. Root development is retarded because of permafrost's chilling effect on soils. As well, roots are forced to grow laterally because downward penetration is prevented by the permafrost. Large trees are not well supported by these shallow root systems, resulting in instability, which is compounded by frost heaving in the *active layer*.

Dyke – A long sheetlike body of *plutonic* rock that fills fractures crossing other rocks. Dykes are differentiated from *sills* in that they cut across the bedding or structural plane of the host rock. American spelling is "dike."

Ecotone – A zone of transition between two or more relatively distinct biological communities. The *taiga*, for example, is an ecotone between the tundra and southern boreal forest. An ecotone typically contains many of the species of each of the overlapping communities (e.g., arctic terns and Caspian terns) and sometimes species that are characteristic of the ecotone itself (e.g., northern shrike; gray-cheeked thrush).

Erosion – The removal of *weathered* rock through the action of flowing water, moving ice and wind.

Fault – Fracture in a rock where observable movement has occurred on one side relative to another in response to local or regional tensions in the earth's crust. Movement may occur laterally, creating a "strike fault," or vertically, creating a "dip fault." Faults on the Canadian Shield commonly show evidence of both horizontal and vertical movement.

Fen – More water flows through fens than *bogs*, allowing the growth of a thick carpet of sedges, their characteristic plant cover. Fens also support some mosses, grasses, willow shrubs and occasional tamarack trees.

Flagging – As a result of the abrasive effect of wind-driven snow and ice particles, spruce trees around the tree line or in very exposed sites elsewhere can take on the appearance of a flag pole, with a clump of needles up top and not much else below except a naked trunk or a few growing points on its downwind side.

Frost heaving – The upward or outward movement of the ground surface or objects on or in the ground (like a boulder or a house), caused by the formation of ice in the soil.

Glacial groove – A smooth, deep, relatively straight furrow cut into bedrock by glacial *abrasion*; much larger than a *striation*.

Glacial Lake McConnell – A large *proglacial lake* that joined the basins of Lake Athabasca, Great Slave Lake and Great Bear Lake about 10,000 years ago. Its final demise is thought to have come 8,700 years ago with a catastrophic flood that overtopped the lake's key plug at the Liard River Delta.

Greenhouse effect – The net warming of the earth's atmosphere by the temporary trapping of outgoing infrared radiation, particularly by carbon dioxide.

Greenstones – Slightly metamorphosed igneous rocks that have been altered from their original structure by relatively low temperatures and pressures. The slightly greenish tinge is due to an abundance of actinolite, chlorite and/or epidote.

Ground fire – A type of forest fire that burns combustible organic matter below the litter layer of the forest floor. Also called a "subsurface fire."

Heat sink – A body or region (such as outer space) capable of unlimited absorption of radiant energy.

Herbivore – Any animal that obtains most of its food from plants. Herbivores are primary consumer organisms and form the second trophic level (above plant producers) in a food chain.

Hibernation – A physiological state in which an animal's body temperature plunges to about the level of its surroundings. Its heart rate, respiration and other metabolic functions similarly fall to low levels (e.g., an arctic ground squirrel). "Torpor" is similar, but while in this state, an animal's body temperature drops only slightly (e.g., a black bear).

Hummock – A bulging mound of soil often having a silty or clayey core and showing evidence of movement due to regular frost action.

Hydrothermal solution – A high temperature broth of fluids, gases and dissolved minerals that travels upwards from the earth's warm interior.

Hypsithermal period – The period of pronounced warmth that marked the end of the *Pleistocene epoch.* Though it spanned approximately 10,000 to 6,000 years, the period of greatest warmth varied in different regions depending on latitude and the location of retreating glaciers.

Ice fog – Atmospheric condition common in northern communities during very cold, still winter days. Moist exhaust from vehicles, homes and office buildings forms a thick cloak of suspended ice crystals in these conditions.

Igneous rock – One of three main groups of rocks (besides *metamorphic* and *sedimentary*) that are typically crystalline in appearance. These rocks solidified from molten magma extruded out onto the earth's surface (*volcanic rock*) or intruded below the surface (*plutonic rock*).

Isotherm – A line on a map passing through areas of equal temperature.

Joint – A clean, planar fracture that cuts through a rock mass without moving it.

Kettle depression – A depression in glacial deposits left by a large fragment of glacial ice.

Land breezes – Light, often gusty breezes that blow from the land to a large water body in response to warm air rising over water. These usually occur during summer nights when prevailing winds are light and the land becomes cooler in relation to the adjacent water surface. (See *sea breezes*.)

Lava – Molten rock upon the earth's surface. (See *magma*.)

Layering – An adaptive response of spruce to other plants that compete with the tree for rooting space. Lower branches put down roots where they come into contact with the competing vegetation (e.g. sphagnum moss) and then send a growing stem upward. The result of this form of vegetative reproduction is a distinctive candelabrum-shaped tree. At the tree line this form of reproduction allows spruce to survive well out into the tundra, beyond the point at which sexual reproduction is possible.

Magma – Molten rock beneath the earth's surface from which all *igneous rocks* are formed.

Marsh – Marshes are productive areas usually found along the shore of a river or lake, which gives them a typically linear shape. Their most characteristic plants include cattails, horsetails and other reedy species. Marshes are subject to wide fluctuations in water levels and may dry out completely by late summer.

Mesozoic era – Literally meaning "middle life," this geological era ranged in time from 240 to 65 million years ago. It was preceded by the *Palaeozoic era* and followed by the *Cenozoic era*.

Metamorphic rock – Rock formed from igneous or sedimentary rocks that have been subjected to great changes in temperature, pressure and/or chemical environment.

Midden – The heap of cone scales discarded by squirrels during feeding. Once a midden becomes thick enough to provide insulation against the cold, squirrels build tunnels, nests and storage chambers within it.

Migration corridor – A mapped expression of the direction of passage and the geographic distribution of waterfowl between breeding areas and wintering areas. (Not to be confused with "flyways," which show very general migration patterns only.)

Moult – A gradual process during which a bird sheds one set of feathers and grows a new set.

Mud boil – A round patch of bare soil created by frost heaving so intense that plants cannot take root.

Palaeocurrents – Curved patterns in *sedimentary rock* that reflect the prevailing wave action or flow of ancient water bodies.

Palaeoeskimos – General term given to prehistoric arctic peoples who migrated into the Canadian Arctic from the Bering Sea region around 4,000 years ago.

Palaeoindians – General term applied to the first known people to inhabit North America following the final *Pleistocene* glaciation. The first material evidence of these people is the distinctive "Clovis" spearheads that date back 11,500 years.

Palaeozoic era – Beginning at the end of the *Proterozoic eon*, this geological period is marked by a relatively rapid proliferation of marine life, hence its literal meaning, "ancient life." This era ranged from 570 to 240 million years ago, ending when the *Mesozoic era* began.

Palsa – A large, peaty permafrost mound having a core of alternating layers of ice and peat.

Peat – A dark brown mass of partially decomposed plant material formed under oxygen-poor conditions in a waterlogged environment. In northern wetlands, peat can cover vast areas of land and be tens of metres deep. Peat has an average carbon content of 50 percent and is used as a domestic fuel in many parts of the circumpolar world.

Permafrost – Soil or rock that remains at or below 0°C for at least two years.

Phytophages – Soil organisms such as nematodes and beetles that eat plant matter.

Pillows (see *Volcanic pillows.*)

Plate tectonics – Motion of the continental plates due to convection currents in the molten mantle layer below the surface.

Pleistocene epoch – The geological time period during the most recent glacial age. It began 1.5 to 2 million years ago and ended approximately 10,000 years ago.

Pleistocene megafauna – Large, extinct mammals that lived during the *Pleistocene epoch.*

Plucking – A process of glacial erosion in which sizable blocks of rock are loosened, picked up and carried away by glaciers. Also called "quarrying." (See *roche moutonée* and *whaleback.*)

Plutonic rock – A general term for any large-scale mass of coarse-grained igneous rock that intruded deep below the earth's surface and has been exposed through millions of years of *weathering* and *erosion.* Also simply called "plutons." Examples include *dykes, sills* and *stocks.*

Pond – A form of wetland characterized by well-defined basins up to several hectares in size filled with still or slow-moving water. Most ponds are shallow, with an average depth of less than four metres. Fed by rainwater, snowmelt or small streams, many ponds owe their existence to the handiwork of beavers. Yellow pond lilies may cover their surface. Cattails, horsetails and other reedy plants often fringe their shores.

Precambrian shield – Shield bedrock (e.g., Canadian Shield) formed during the *Archean* or *Proterozoic* eons.

Proglacial lake – A short-lived, constantly shifting water body formed from meltwaters at the trailing edge of a retreating glacier, for example, *Glacial Lake McConnell. Pro* is Latin for "before;" hence it is literally a "lake before the glacier."

Proterozoic eon – The period measured on the geological time scale that spans from the end of the *Archean eon,* 2.5 billion years ago, to the beginning of the *Palaeozoic era,* 570 million years ago. The Archean and Proterozoic eons together make up the *Precambrian era.*

Quartz vein – A tablelike or sheetlike body of quartz that has intruded into a joint or fissure in rocks.

Refection – The reingestion of incompletely digested faecal pellets, rechewing, and ultimate digestion, as practised by hares and rabbits.

Roche moutonée – Asymmetrical, glacially eroded bedrock knob with a smooth, more gentle flank on the upstream side (*abraded*) and steep, irregular, jagged flank on the downstream side (*plucked*).

Rock drumlin – A streamlined rock knob sculptured by glacial abrasion. It looks much like a *whaleback* but is much longer — up to 100 metres — and has a steeper, blunter side facing upstream and a gentler, narrower tail facing downstream. Seen from above, a rock drumlin has the shape of an elongated teardrop.

Saprophages – Soil organisms that eat dead and decaying matter. These include millipedes, mites, bacteria and fungi.

Sea breezes – Light breezes that blow from a large water body to the adjacent land in response to strong daytime heating over the land surface. These usually occur during summer days when prevailing winds are light and the land becomes warmer in relation to the adjacent water surface. (See *land breezes.*)

Sedimentary rock – Rocks formed by the accumulation and cementing of sediments.

Shield country – A term used in this book as synonymous with the western portion of the *taiga shield ecozone.*

Silicates – Of some 2,000 different minerals known to be present in the earth's crust, the ones making up the majority of more common rocks are the silicates. Though these minerals vary in colour, texture and hardness, they all include some kind of metal combined with silicon (hence the name) and oxygen. Common silicates of the Canadian Shield include feldspar (orthoclase and plagioclase), quartz, mica (biotite and muscovite), pyroxenes and amphiboles.

Silicified – Describes rocks that have been hardened by replacement or infilling of large amounts of quartz or silica.

Sill – A sheetlike body of *plutonic rock* that originally intruded into fractures that run parallel to the bedding or structural plane of the host rock. (See *dyke* for comparison.)

Slave geological province – One of seven geological (or "structural") provinces of the Canadian Shield. Situated northeast of Great Slave Lake, this province contains the oldest known rocks on earth, aged at 3.962 billion years old. (Recent advances in dating technology suggest that they may be over 4 billion years old.) The Slave province is 190,000 square kilometres in total area — a little larger than the island of Labrador.

Snag – A standing dead tree from which at least the leaves or needles and smaller branches have fallen off or been torched by a forest fire. Often only the main trunk is left standing.

Stock – A roughly circular or oval-shaped body of *igneous rock* that increases in size with depth, has no known floor and has an exposed surface area of less than 100 kms².

Striation – Small linear groove in rock. The term is usually applied to parallel surface grooves formed by glacial *abrasion.*

Stromatolite – Precambrian fossils of layered rock formed by toadstool-like colonies of blue-green algae (also called cyanobacteria). These organisms were the dominant life form on earth between about 3 billion years ago and 500 million years ago.

Sundog – An atmospheric phenomenon in which a halo of light appears around the sun or as isolated patches of brightness on both sides. Also called a "parhelion" or "mock sun," this effect is created most often in winter by the refraction of low-angled sunlight through long, narrow ice crystals in cirrus or cirrostratus clouds.

Surface fire – A type of forest fire — the most common one — that burns all combustible materials lying above the ground but below the tree canopy.

Swamp – A swamp is simply a *marsh* that supports trees and tall shrubs. Still or gently flowing water covers most of the land surface during the wetter periods of the year. Swamps are found most commonly along major rivers in the Western Arctic. Typical plant species include balsam poplar trees and alder and willow shrubs.

Symbiosis – Biological partnership entered into by two unrelated organisms, each of which has something the other needs.

Taiga – Pronounced "TIE-gah," this Russian word originally referred to a dense, marshy forest in Siberia. It now is used more generally to refer to the northern boreal forest stretching across the entire circumpolar world.

Taiga shield ecozone – One of 15 terrestrial ecozones of Canada identified by Environment Canada (1986). It occurs where the *taiga* forest and *Precambrian shield* overlap. This ecozone stretches across the upper two-thirds of the continent from the Labrador coast in the east to the shores of Great Bear Lake in the west.

Tailings – Waste rock fragments produced as a result of processing ore in a mine.

Tussock – A narrow hump of sedges and peat often occurring in large numbers in wetland habitats.

Veteran spruce – A spruce tree that is usually much taller and older than the surrounding vegetation. Often found in small, isolated groups, they represent survivors of a previous forest fire that may have burned most of the other trees around them. This effect is occasionally also demonstrated by large pine trees.

Volcanic pillow – Distorted globular masses formed from *lava* that spilled out onto the ocean floor under tremendous pressure of water from above. Also called "pillow lava."

Volcanic rocks – Rocks that formed by the extrusion of *lava* onto the earth's surface.

Weathering – The breakdown and decay of rocks through physical or chemical means. When rain falls on the surface of rocks, it percolates into tiny pores and cracks where it may later freeze, expand and force apart individual grains or large chunks of rock. Other physical weathering agents include plant roots and the natural swelling and contraction of various minerals. Common chemical weathering agents include oxygen, carbon dioxide, rainwater and various compounds released by rotting plant matter. (See *erosion*.)

Western Plutonic Complex – A massive granitoid *pluton* situated in the southwest corner of the *Slave geological province*.

Wetland – Any area of generally low, flat land that holds water during at least part of the year. This water may be at, near or above the ground surface.

Whaleback – Smooth, glacially sculptured bedrock knob shaped and sized like the back of a sounding whale. (See *abrasion*.)

Wildfire – An unplanned or unwanted natural or human-caused fire, as opposed to a "prescribed fire," which is deliberately set and controlled for the purposes of forest management.

Wind chill – A simple measure of the chilling effect experienced by the human body when strong winds are combined with freezing temperatures. The wind chill temperature or heat loss factor is a good indicator of what one should wear to be protected from the cold. Two means of expressing wind chill are heat loss in watts per square metre (W/m^2) and the equivalent temperature in °C.

Wind slab – Upper layer of snow that has been hardened by the breaking up and recrystallization of snow crystals moved by the wind.

References

ANDRUSS, V.C., PLANT, J., and WRIGHT, E., eds. 1990. Home! A Bioregional Reader. Philadelphia: New Society Publishers.

ATMOSPHERIC ENVIRONMENT SERVICE. 1986. The Reliability of Weather Forecasts, Fact Sheet. Downsview, Ontario: Environment Canada.

_____. 1993. Wind Chill Chart. (This appears regularly on the weather page of *News North* during the winter months.) Yellowknife: Northern News Service.

BARR, W. 1991. Back from the Brink: The Road to Muskox Conservation in the Northwest Territories. Calgary: Arctic Institute of North America.

BASTEDO, J.D. 1986. An ABC Resource Survey Method for Environmentally Significant Areas in Canada's North. Waterloo: University of Waterloo, Department of Geography. Publication Series No. 24.

BELL, J.M. 1902. On the Topography and Geology of Great Bear Lake and Thence to Great Slave Lake. Ottawa: Geological Survey of Canada. GSC Annual Report 12(Part C):5–28.

BERGERUD, A.T. 1974. The Decline of Caribou in North America Following Settlement. Journal of Wildlife Management 38(4):757–770.

BIRD, J.B. 1972. The Natural Landscapes of Canada — A Study of Regional Earth Science. Toronto: Wiley Publishers of Canada Ltd.

BOWRING, S.A., WILLIAMS, I.S., and COMPSTON, W. 1989. 3.96 Ga Gneisses from the Slave Province Northwest Territories, Canada. Geology 17:971–975.

BROMLEY, M. 1986. Fur Trade in the Northwest Territories — From Earliest Days to the Present Time. In: Hall, E., ed. A Way of Life. Yellowknife: Government of Northwest Territories, Department of Renewable Resources.

BROPHY, J. 1987. Origin of Lakes Is a Legacy of the Last Ice Age. News North, October 26:11.

BROWN, R.J. 1970. Permafrost as an Ecological Factor in the Subarctic. Ecology of Subarctic Regions — Proceedings of the Helsinki Symposium. Paris: UNESCO.

BURY, H.J. 1917. Conservation of Timber in the Province of Alberta and the Northwest Territories. Report cited in Janzen, 1990.

CAMSELL, C. 1954. Son of the North. Toronto: Ryerson Press.

CANADIAN ARCTIC RESOURCES COMMITTEE. 1987. A Question of Degrees. Northern Perspectives 15(5).

CRAIG, B.G. 1965. Glacial Lake McConnell, and the Surficial Geology of Parts of Slave River and Redstone River Map Areas, District of Mackenzie. Ottawa: Geological Survey of Canada. Bulletin No. 122.

DAVIS, W. 1993. Shadows in the Sun: Essays on the Spirit of Place. Edmonton: Lone Pine Publishing.

ENVIRONMENT CANADA. 1986. Terrestrial Ecozones of Canada. Ottawa: Supply and Services Canada.

_____. 1989. Depletion of the Arctic Ozone Layer. Changing Atmosphere Fact Sheet. Ottawa: Supply and Services Canada.

_____. 1990. The Climates of Canada. Ottawa: Supply and Services Canada.

_____. 1991. Marine Guide to Local Conditions and Forecasts. Ottawa: Supply and Services Canada.

ERSKINE, A.J. 1977. Birds in Boreal Canada — Communities, Densities and Adaptations. Ottawa: Supply and Services Canada. Canadian Wildlife Service Report Series No. 41.

FINNIE, R. 1948. Canada Moves North. Toronto: The MacMillan Company.

FRANKLIN, J. 1824. Narrative of a Journey to the Shores of the Polar Sea, in the years 1819–20–21–22. London: John Murray.

FREDEN, G.H. 1990. Pike's Portage: Gateway to the Barrens. Up Here, January/February: 71–72.

FUMOLEAU, R. 1973. As Long as This Land Shall Last. Toronto: McClelland and Stewart.

GANDHI, M.K. 1982. All Men Are Brothers, Autobiographical Reflections. New York: Continuum Publishing Corporation.

GARRETT, W.E. 1988. The Peopling of the Earth. National Geographic 174(4):434–503.

GODFREY, W.E. 1986. The Birds of Canada. Ottawa: National Museums of Canada.

GORE, R. 1989. The March toward Extinction. National Geographic 175(6):662–699.

GOULD, G., ed. 1984. Jock McMeekan's Yellowknife Blade. Duncan, British Columbia: Lambrecht Publications.

GOVERNMENT OF THE NORTHWEST TERRITORIES. 1976. Book of the Dene. Yellowknife: Department of Education, G.N.W.T.

_____. 1985. Canoe trip reports for various rivers in the Northwest Territories. Unpublished files retained by the Department of Economic Development and Tourism in Yellowknife.

_____. 1991. The Flames of Regeneration. Video. Yellowknife: Department of Renewable Resources.

_____. 1993a. Dene Kede: Education — A Dene Perspective. Yellowknife: Department of Education, Culture and Employment.

_____. 1993b. An Investigation of Atmospheric Emissions from the Royal Oak Giant Yellowknife Mine. Yellowknife: Department of Renewable Resources.

GRAVES, J. 1988. A History of Wildlife Management in the Northwest Territories. Report prepared for the Department of Renewable Resources, Yellowknife.

HALE, M.E. 1979. How to Know the Lichens. Dubuque, Iowa: Wm. C. Brown Company Publishers.

HENDERSON, J.F., and JOLLIFFE, A.W. 1939. Relation of Gold Deposits to Structure, Yellowknife and Gordon Lake Areas, Northwest Territories. Ottawa: The Canadian Institute of Mining and Metallurgy, Geological Survey of Canada.

HENRY, J.D. 1988. Taiga — Earth's Evergreen Mantle. Borealis 1(1):6–12.

HEWITT, G. 1916. The Conservation of Our Northern Mammals. Report prepared for the Commission of Conservation, Ottawa.

HOCKING, D., KUCHAR, P., PLAMBECK, J.A., and SMITH, R.A. 1978. The Impact of Gold Smelter Emissions on Vegetation and Soils of a Sub-Arctic Forest-Tundra Transition Ecosystem. APCA Journal 28(2):133–137.

INDIAN AND NORTHERN AFFAIRS CANADA. 1973. Forest Fire Reports for the 1973 Season, Unpublished archive files retained in the Government of Northwest Territories' Fire Centre in Fort Smith.

IRELAND, R.R., BIRD, C.D., BRASSARD, G.R., SCHOFIELD, W.B., and VITT, D.H. 1980. Checklist of the Mosses of Canada. Ottawa: National Museums of Canada, National Museum of Natural Sciences. Publications in Botany No. 8.

JANZEN, S.S. 1990. The Burning North — A History of Fire and Fire Protection in the Northwest Territories. M.A. thesis, Department of History, University of Alberta.

KELSALL, J.P. 1968. The Migratory Barren-Ground Caribou of Canada. Ottawa: Department of Indian Affairs and Northern Development.

KITTO, F.H. 1920. Report of a Preliminary Investigation of the Natural Resources of Mackenzie District. Ottawa: typescript.

LaCHAPELLE, E.R. 1992. Field Guide to Snow Crystals. Cambridge: International Glaciological Society.

LARSEN, J.A. 1980. The Boreal Ecosystem. New York: Academic Press Inc.

LAURIAULT, J. 1989. Identification Guide to the Trees of Canada. Markham, Ontario: Fitzhenry and Whiteside.

LAYTHA, E. 1939. North Again for Gold. New York: Frederick A. Stokes Company.

LECHOWICZ, M.J. 1982. The Effects of Simulated Acid Precipitation on Photosynthesis in the Caribou Lichen. Water, Air and Soil Pollution 18:421–430.

LEMONICK, M.D. 1992. The Ozone Vanishes. Time, 17 February:40–43.

LEVENSON, T. 1989. Ice Time — Climate, Science & Life on Earth. New York: Harper & Row.

LOPEZ, B. 1987. Arctic Dreams. New York: Bantam Books.

LOWER, A.R.M. 1936. Settlement and the Forest Frontier in Eastern Canada. In: Mackintosh, W.A., and Joerg, W.L. Settlement and the Forest and Mining Frontiers. Toronto: Macmillan. Cited in Janzen, 1990.

MacDONALD, W. 1942. Report on a Visit to North-West Canada and Alaska. Cited in Janzen, 1990.

McELROY, T.P. 1974. The Habitat Guide to Birding. New York: Alfred A. Knopf.

McFARLANE, R. 1880. McFarlane Papers. Unpublished papers and correspondence of the Church Missionary Society 1897–1901, Vol. 1, Fort Chipewyan, 24 December.

MOON, B. 1970. The Canadian Shield. Toronto: Natural Science of Canada Ltd.

MORELL, V. 1990. Rare Rocks. Equinox 51:26.

NOBLE, W.C. 1981. Prehistory of the Great Slave Lake and Great Bear Lake Region. In: Helm, J., ed. Handbook of North American Indians. Washington, D.C.: Smithsonian Institution. 97–106.

NORTHERN NEWS SERVICE. 1991. Rare Duck Nests in YK Family's Birdhouse. Yellowknifer, 12 July.

_____. 1992. Hole in Ozone Raised Cancer Risk. News North, 10 February.

PADGHAM, W.A. 1987. Yellowknife Guidebook — A Guide to the Geology of the Yellowknife Volcanic Belt and Its Bordering Rocks. Yellowknife: Yellowknife Geo-workshop Committee and Indian and Northern Affairs Canada.

_____. 1990. The Slave Province, an Overview. Mineral Deposits of the Slave Province, Northwest Territories. Field Trip Guidebook for 8th IAGOD Symposium. Ottawa: Geological Survey of Canada. Open file no. 2168:1–40.

_____. 1991. Mineral Deposits in the Archean Slave Structural Province. Yellowknife: Indian and Northern Affairs Canada.

PARKER, J.M. 1971. Emporium of the North. In: Chalmers, J.W., ed. On the Edge of the Shield: Fort Chipewyan and Its Hinterland. Edmonton: Boreal Institute for Northern Studies. Occasional Publication No. 7:14–20.

PIELOU, E.C. 1991. After the Ice Age — The Return of Life to Glaciated North America. Chicago: University of Chicago Press.

PORSILD, A.E., and CODY, W.J. 1980. Vascular Plants of the Continental Northwest Territories, Canada. Ottawa: National Museum of Natural Sciences, National Museums of Canada.

POTTS, G.W. 1988. Pillow Structure of the Yellorex Flows near Yellowknife, Northwest Territories. B.Sc. thesis, Department of Earth Sciences, Carleton University.

PRICE, R. 1967. Yellowknife. Toronto: Peter Martin Associates.

PRUITT, W.O. 1978. Boreal Ecology. London: Edward Arnold Publishers. The Institute of Biology's Studies in Biology No. 91.

_____. 1983. Wild Harmony — The Cycle of Life in the Northern Forest. Saskatoon: Western Producer Prairie Books.

_____. 1989. Address at the Northern Science Award Presentation, 9 November 1989. Canadian Field Naturalist 103(4):606–609.

_____. 1991. Possible Greenhouse-Induced Habitat and Faunal Changes in the Taiga of Central Canada. Paper presented at the symposium, Climate Change and the Boreal Forest. Annual general meeting of the Canadian Society of Zoologists, Lakehead University, Thunder Bay, Ontario, 8–11 May 1991.

RAFFAN, J., and SIMMONS, G.C. 1980. The Burnside River, N.W.T.: Its Environ's Natural History and Potential for Recreational and Interpretive Experience. Ottawa: Parks Canada.

REDFERN, R. 1983. The Making of a Continent. New York: Times Books.

RICH, E.E., ed. 1968. Simpson's Athabasca Journal. In: The Publications of the Hudson's Bay Record Society. Vol. 1. Nendeln, Liechtenstein: Klaus Reprint Limited. 371–372.

RITCHIE, J.C., and HARE, F. 1980. Late-Quaternary Vegetation and Climate near the Arctic Tree Line of Northwestern North America. Quaternary Research 1:331–342.

ROBERTS, E., and AMIDON, E. 1991. Earth Prayers from Around the World. New York: Harper and Collins.

ROGERS, E.S., and SMITH, J.G.E. 1981. Environment and Culture in the Shield and Mackenzie Borderlands. In: Helm, J., ed. Handbook of North American Indians. Washington, D.C.: Smithsonian Institution. 130–145.

RUSSELL, F. 1898. Explorations in the Far North. Des Moines: University of Iowa.

SALE, K. 1983. Mother of All — An Introduction to Bioregionalism. Great Barrington, MA: E.F. Schumacher Society.

SCHINDLER, D.W., BEATY, K.G., FEE, E.J., CRUIKSHANK, D.R., DEBRUYN, E.R., FINDLAY, D.L., LINSEY, G.A., SHEARER, J.A., STAINTON, M.P., and TURNER, M.A. 1990. Effects of Climatic Warming on Lakes of the Central Boreal Forest. Science, 16 November:967–970.

SCOTTER, G.W. 1966. A Contribution to the Flora of the Eastern Arm of Great Slave Lake, Northwest Territories. The Canadian Field-Naturalist 80(1):1–18.

_____. 1977. Fire and Caribou in Northern Canada. In: Dube, D.E., compiler. Fire Ecology and Resource Management — Workshop Proceedings. Edmonton: Environment Canada.

SMITH, D.G., ed. 1981. The Cambridge Encyclopedia of Earth Sciences. Scarborough: Prentice Hall Canada Inc.

STONEHOUSE, B. 1971. Animals of the Arctic — The Ecology of the Far North. New York: Holt, Rinehart and Winston.

TARNOCAI, C. 1978. Distribution of Soils in Northern Canada and Parameters Affecting Their Utilization. In: 11th International Congress of Soil Science Transactions — Symposia Papers. Edmonton: International Society of Soil Science. 281–304.

THE PRESS INDEPENDENT. 1991. Trees Are Threatened — Spruce Budworm Spreading. 9 August. Yellowknife.

THIERET, J.W. 1964. Botanical Survey along the Yellowknife Highway, Northwest Territories, Canada. SIDA 1(4):187–239.

THOMSON, J.W., SCOTTER, G.W., and AHTI, T. 1969. Lichens of the Great Slave Lake Region, Northwest Territories, Canada. The Bryologist 72:137–177.

THURSTON, H. 1986. Icebound Eden. Equinox 27(May/June):72–85.

VANDERBURGH, S., and SMITH, D.G. 1988. Slave River Delta: Geomorphology, Sedimentology, and Holocene Reconstruction. Canadian Journal of Earth Sciences 25(12):1990–2004.

WALKER, M. 1984. Harvesting the Northern Wild. Yellowknife: Outcrop Ltd.

WEINER, J. 1990. The Next Hundred Years — Shaping the Fate of the Living Earth. New York: Bantam Books.

WRIGHT, J.V. 1981. Prehistory of the Canadian Shield. In: Helm, J., ed. Handbook of North American Indians. Washington, D.C.: Smithsonian Institution. 86–96.

Index

About the Author

Jamie Bastedo has a Master of Arts in Regional Planning and Resource Development from the University of Waterloo. He has lived and worked in the North for more than fifteen years. For several of those years he was the popular host of CBC Radio's Northern Nature series, broadcast live and always outdoors from shield country's subarctic environs. In 1990, Jamie Bastedo launched CYGNUS Environmental Consulting, which specializes in environmental planning, assessment, education and eco-tourism. His previous books include *Blue Lake and Rocky Shore* and *Reaching North: A Celebration of the Subarctic*. Jamie Bastedo lives in Yellowknife with his wife and two daughters.

Also by Jamie Bastedo

"The land of little sticks, the land of bedrock and bush, the land of a million lakes . . ." The clichés pegged to Canada's subarctic are as rich and seemingly endless as the landscape they describe. In *Reaching North: A Celebration of the Subarctic,* Jamie Bastedo looks beyond those clichés into the heart of one of Canada's largest yet least understood regions—the subarctic wilderness. Follow the complete life cycle of a tiny snow crystal and a bedrock fault a thousand kilometers long. Enter into a frozen beaver lodge in January and the heart of a midsummer fire storm. Ponder the delicate hues of a raven's egg and the riot of color that is the northern lights. Told with a passion for northern people and an exacting eye for detail, *Reaching North* celebrates a vast northern wilderness, drawing inspiration from its basement bedrock all the way to the aurora borealis.

ISBN 0-88995-170-5 • paperback
256 pages • CAN 16.95 • USA 14.95

Praise for *Reaching North: A Celebration of the Subarctic*

"Bastedo has an effortless expertise on seemingly all aspects of northern life, and his essays freely and confidently range between colourful personal anecdotes and lucid scientific explanations of various natural phenomenon. . . . *Reaching North* is an absolute pleasure to read." *–Vue*

"Bastedo, a writer with a gift for explaining the mysteries of the natural world, clearly loves the land where he lives and the people he finds there. His enthusiasm is infectious." *–Yellowknifer*

"Bastedo strays from the well-traveled routes of history, memoir and science to write instead of people. He takes us inside their powerful emotions and let's us share their feelings and experiences. . . . Inspiring." *–Up Here*

"A thoroughly enjoyable read." *–CBC North*